CHESAPEAKE COLLEGE
THE LIBRARY
WYE MILLS,
MARYLAND 21679

DATE	ISSUED TO

© DEMCO 32-2125

THE INNOVATION FORMULA

THE INNOVATION FORMULA

How Organizations Turn Change into Opportunity

Michel Robert
Alan Weiss

Ballinger Publishing Company
Cambridge, Massachusetts
A Subsidiary of Harper & Row, Publishers, Inc.

Copyright © 1988 by Ballinger Publishing Company. All rights reserved. No part of this publication may be reproduced, stored in a retrieval system, or transmitted in any form or by any means, electronic, mechanical, photocopy, recording or otherwise, without the prior written consent of the publisher.

International Standard Book Number: 0–88730–352–8

Library of Congress Catalog Card Number: 88–10578

Printed in the United States of America

Library of Congress Cataloging-in-Publication Data

Robert, Michel, 1941–
The innovation formula.

Includes bibliographical references and index.
1. Organizational change. 2. Entrepreneurship.
I. Weiss, Alan Jay, 1946– . II. Title.
HD58.8.R63 1988 658.4′06 88–10578
ISBN 0-88730–352–8

CONTENTS

057599

LIST OF FIGURES

ACKNOWLEDGMENTS

We owe a tremendous debt to the clients of Decision Processes International, whose organizations were the "laboratories" in which our observations and validations of the process of innovation were conducted. They know the approach as Entreprennovation™ and we thank them for their continuing feedback about innovative methodology.

Our thanks to the Decision Processes International partners around the world. They kept us focused on practical results to be achieved in actual client environments. We are indebted to them for their ability to prevent cultural bias and untested ideas from entering the process.

Finally, for Ellie and Marie, here is the result of those horrible travel schedules and missed dinners. . . .

INTRODUCTION

Entrepreneurs are romantic figures. They are viewed as business swashbucklers who catapult new ideas into public prominence while they storm the walls of the establishment. This is great adventure. It's also pure fiction.

Our study of entrepreneurs paints quite a different picture. We've examined the famous, such as Ray Kroc of McDonald's and Fred Smith of Federal Express, and the not so famous, such as the middle managers who have accounted for decades of successful innovation at Merck and 3M and Bell Labs. And we've found that true entrepreneurs aren't pirates, but disciplined sailors who anticipate the winds and tides of change. Entrepreneurship isn't a matter of drawing to an inside straight, but one of stacking the deck from the start.

We've long suspected that there is a process at work with successful entrepreneurs, a process that many can't—or won't—articulate. We've also felt that the process has to be fairly simple and straightforward, one that can be used both individually and organizationally, without complexity. Our research has supported these beliefs. We have worked with top management in hundreds of firms on six continents over the past twenty years, and we're convinced that the next ten years will be the age of entrepreneurial management.

And we're not alone. There are scores of books, audio/video tapes, and courses that reinforce this view. So why add yet another book to the mass? Because, up to this point, no one has provided managers with a "how-to." Most books are written first, with the intent that organizations then follow the advice. We studied successful organizations first, then put the processes we found in a book. It's one thing to study someone else's successes, but it's generally useless, once you change the environment and the people, to try to emulate what was done elsewhere. Our approach is one of process, not content. We don't seek to catch a fish for you, but rather to teach you *how* to fish, so that you can apply the principles in your particular pond. Our interest is in the techniques—the systematic, learnable techniques—that have been successfully employed. In isolating and codifying these approaches, we can provide a methodology for entrepreneurship.

We want to make clear at the outset just what the relationship between entrepreneurs and innovation encompasses. Entrepreneurs innovate. Noun and verb. Object and action. As simple as that. This book is about how they do it, and about how *you* can do it today, tomorrow, at work, at home. It is based on observation, development of concepts, implementation, and validation within client organizations.

The United States has traditionally been a world leader in technology, economic strength, financial clout, and, generally, "getting things done." We are a nation of overachievers. We believe that these achievements have been directly correlated to the entrepreneurial spirit of the country. The degree to which we suffer in any of these areas today is, similarly, a result of a decline of entrepreneurialism—that is, the willingness to innovate. The old trial-and-error, "damn the torpedoes" approaches to innovation no longer serve us well in an increasingly competitive global economy. Years ago, marketing an unsuccessful new product meant that you went back to the drawing board. Today, it often means Chapter 11. People Express innovated itself right out of existence. New Coke became the Edsel of the 1980s. General Motors' retreat from innovation in the name of cost-consciousness created a company of look-alike cars that decimated its luxury auto sales. Yet the *Wall Street Journal* can appear in the mail of its nationwide readership every day, and Merck & Company can consistently produce new drugs in areas where others have failed.

The entrepreneurial organization is not a passing fad that will go the way of T-groups, transactional analysis, and quadraphonic sound. All of us in organizational structures will be repeatedly called upon to be innovative in solving problems, making decisions, and planning our futures. We present here our method for being entrepreneurial in the true sense: viewing change as opportunity, not threat, and using a process to manage that change so as to exploit it in your organization's best interests. That is the heart and soul of innovation. In the following pages, we present the art and science.

—Michel Robert
—Alan Weiss

Chapter 1
THE AGE OF ENTREPRENEURIAL MANAGEMENT

All around us, the structure and fabric of business are changing. Just as the rampant inflation of the late 1970s forever changed our economy and society, the technological, economic, and demographic changes of the 1980s have permanently changed fundamental management practices. Many firms have thrived on the changes—organizations like McDonald's, Merck, and Federal Express. In other cases, entirely new firms and industries have arisen as a result of change—organizations such as health maintenance organizations, video rental stores, and franchise operations ranging from printing to hardware to cookies. Why do some firms prosper and grow in the midst of change, while others deteriorate and decline?

We believe the answer lies in the ability to manage change. This ability stems from accepting change as opportunity, not threat. But most people don't understand how to do this as a repeatable business practice. They tend to shy away from trying to exploit change, as if it's some risky business that's best left to the shoot-from-the-hip, wild-eyed entrepreneur who will bet everything on a roll of the dice. Nothing could be further from the truth. We define an *entrepreneur* as someone who redeploys assets and resources from areas of low productivity and yield to areas of high productivity and yield. Entrepreneurs achieve this redeployment by *innovating,* which consists of the systematic anticipation, recognition, and exploitation of change.

Change is the basic fuel of innovation, its source, its raw material. The process of innovation is the tool of the entrepreneur. These are the definitions we'll be using throughout this book.

Innovation has rapidly assumed a position of prominence in the business community. This is because managing the future, staying ahead of the competition, keeping abreast of new developments, and the management of marketing, selling, and servicing in the best possible way in a variety of circumstances are the key strengths of any organization hoping to prosper over the next decade. This is, of course, easier said than done. Let's look at just one example.

THE THREAT OF COMPETITION

Years ago, Friden had a dominant position in mechanical calculators. Walking into any large company, from Prudential Insurance to General Motors, the visitor was likely to see hundreds of Friden calculators planted on acres of desks. In retrospect, they were ungainly, noisy, and slow, but they were all there was for the job. With the advent of electronic technology, however, mechanical calculators became obsolete within a very brief time span. Some firms viewed the new technology as an opportunity and promptly set about adapting it to the needs of desktop calculation. But others saw the new technology as a threat and tried to improve the existing mechanical technology to combat the threat. Today, we defy you to find a contemporary business operating with Friden calculators of any type. Friden is out of the business because Friden management didn't have a *process* for viewing change as opportunity. They could only react, combating change with old ideas.

This is not an isolated instance. By and large, the manufacturers of vacuum tubes abandoned the field to transistors. The Swiss watch industry, the longtime expert in manufacturing analog watches, had the rug pulled out from under it by quartz and digital technology. Synthetics turned the American cotton industry upside down. One of the primary requirements for successful innovation is being receptive to change, not resistant to it. This simple, often unconscious, mental attitude can make all the difference. It did for Friden.

Our projection is that the traditional, shortsighted style of management—"defending its turf" and creating fiefdoms of power—must become obsolete. Change is too complex and wide-ranging for such keyhole views. Managers will have to cooperate to be able to view change holistically and to understand what opportunities may await. Hence the need for a process that is objective and utilizes information from a variety of sources. One reason for combining as many good minds as possible in this pursuit is that innovation demands that all eyes be kept on the *result,* not on the managerial activity or on a particular product or service. A famous management aphorism is that the drill was not invented because anyone needed a drill but because people needed holes, and the drill was merely the best way of providing them. Similarly, customers don't need their parcels delivered by jet or their hotel rooms locked by electronic cards or their favorite television programs taped. But they *do* want next-day delivery of important mail, security in hotels, and the flexibility to time-shift television programs for later viewing. Successful innovators concentrate on the end result—the effect on the business environment—rather than on the product.

Customers, competitors, suppliers, agents, government—all of these sources and others need to be thoroughly scrutinized for opportunity. Such scrutiny must be systemized and institutionalized, since the process of innovation is no good as a one-time accomplishment. Innovation must be no less ongoing and routine than quality control or financial review. The one-time innovators, whether Osborne Computer or People Express, seldom sustain their initial growth, and sometimes fail altogether. It is the Bell Labs, the retailers like Marks and Spencer, the Mercks, and franchisers like McDonald's—the consistent, unrelenting innovators—who remain ahead of the pack.

New Competitors

Competition used to be United Airlines versus American Airlines, each offering the same regulated fares. Or Macy's versus Gimbel's. Or Ford and Chevrolet, Coca-Cola and Pepsi. Today, competition is exacerbated by technology, government intervention or lack of intervention, lifestyles, perceptions, growth, changing industry structure, new knowledge, demographics, and a host of other variables. United and American now compete in an unregulated environment against new, low-cost carriers at a time when consumer decisions are often made on the basis of frequent-flyer bonus points. Gimbel's is gone, and Macy's competes against everything from other large chains such as K Mart and Bloomingdale's to small boutiques offering personalized service—even against the local pharmacy, which has become a mini-department store right down the block. Ford and Chevy are now part of an industry in which about 170 different car models are produced annually in the United States alone, and in which imports are at an all-time high in a society that has moved from fuel-consciousness to safety-consciousness to status-consciousness.

Competition, because of change, is becoming harder to anticipate, harder to track, harder to understand, and much harder to combat. Competition no longer simply means traditional competitors. It now means *any*thing—anyone, any organization, any movement—that takes your customers' money away from you. So movie theaters now compete against home-video rental centers and cable television; the U.S. Postal Service competes against Federal Express, and both compete against electronic mail; your bank competes against Merrill Lynch and against Sears, all of them offering a variety of competing financial services; long-distance bus companies compete against airlines—offering no-frill fares, of all things—and both compete against Amtrak; board games compete against video games and computer games; mail-order catalogues compete against computer shopping; travel agents compete against direct, desktop-computer reservations systems. How is one to stay alive, much less dominate? The secret is innovation, and it's being done successfully all around us.

Movie theaters have cleaned up their act by offering ten or more movies in one location, installing comfortable seats and dynamic sound systems, and attempting to maintain a clean, family atmosphere. They make substantial profits from food purchases and have expanded their food and beverage offerings. Many have even begun displaying art works for sale, an ideal marketing opportunity since many people waiting for the show are staring at the walls anyway. Similarly, the post office has instituted express mail. The variable rate mortgage, cash reserve accounts, debit cards, free travel insurance, baggage insurance, and other new offerings all resulted from the competition described above. You must view competition in terms of finding *alternative ways for the consumer to obtain the results desired,* not as trying to underprice or outperform specific companies within your industry. While the latter kind of competition will always exist, it is just one component of the overall competitive picture.

We believe that the systematic pursuit of opportunities will enable you to more effectively compete. It will make you proactive and, therefore, less vulnerable to the moves of the competition or the marketplace. If you continue to systematically innovate, the competition will have to *react to you,* and that is the secret of market dominance. Just ask IBM.

WINNERS AND LOSERS

Innovation is typically seen as a solitary undertaking, requiring boldness, creativity, and, perhaps, brilliance. Our view is that innovation is an organizational preparedness for accelerating change and for widespread, new sources of competition. Moreover, it is an *offensive* weapon much more than a defensive one. Henry Ford, one of the great entrepreneurs, said, "It could almost be written down as a formula that when a man begins to think that he at last has found the method, he had better begin a most searching examination of himself to see whether some part of his brain has not gone to sleep."[1] Innovation needs to be constant. Too many people and organizations have "gone to sleep" when they thought they had a lock on their marketplace. *Innovation is, and should be, a repeatable business practice.*

Organizations such as IBM, Hewlett-Packard, Johnson & Johnson, and others have demonstrated that innovation should be an essential part of management practice. They have seen that change is inevitable, opportunity is plentiful, and a process is required to manage it. And they have tended to use innovation as an *offensive* weapon. They take prudent risks when innovating. But they seem to know that *not* innovating is the riskiest course of action of all.

One of the key reasons for making innovation an ongoing pursuit is that almost every product, service, and approach in the marketplace ultimately reaches a mature stage and declines. (Or plateaus, at best; but even a plateau is dangerous because level sales do not generate the funds necessary to provide for sustained growth.) As mentioned earlier, vacuum tubes were perfected, but then a sudden leap in technology replaced them with transistors and semiconductors. Mechanical calculators were still being refined, when electronics made them obsolete. In neither case could further research or improvements make the old technology competitive with the new. Why did management continue to try to improve the old? Because they were content to sit on their success—to be complacent with their market—and were defensively oriented. They wanted to protect, not to exploit.

It is essential that organizations know their limits—that is, that they understand the farthest point to which existing technology and approaches are capable of competitive improvement. They should be investigating alternative methods—innovating—before the plateau is reached, and certainly before a new technology makes their existing one obsolete. Such "gaps"—radical transitions from existing approaches to dramatically improved ones—are called "discontinuities" by authorities like Peter F. Drucker and Richard Foster.[2] These discontinuities will undoubtedly accelerate as the century draws to a close. We believe that innovative companies and entrepreneurial management will best bridge the gaps and will be the first to see opportunities.

Down With the Ship

Many companies actually have strategies and top-management policies that inhibit any attempt at innovation. You've heard of them. "Our core belief is that we must respect the integrity of our basic business and protect it at all costs." "Each of our products and services is unique, and we will strive to protect them from competitive inroads." Whether a computer firm, a training company, a bank, or a travel agency, no organization will long prosper if the direction from above is to protect rather than to innovate; if the attitude is, "If we're careful, things won't change," rather than expecting that things *must* always change; if the expectations are for increased sales based on constantly improving the product rather than for breakthrough sales based on the constant search for *better* technologies and better sales approaches.

It is noble for captains to go down with the ship, one supposes, but not very practical. It is neither noble nor intelligent for managers to go down with the business. But managers who can't accept change will inevitably do so. We are not talking about small or little-known organizations, though these, too, are vulnerable. Virtually *all* companies are struggling with changing times and with innovation. But why does Piedmont succeed, when Eastern fails? Why do Merck and Johnson & Johnson excel in their industry? Because some managers and some organizations are much better equipped to manage change and to take an offensive, innovative approach to changing times. One indication of the lack of attention paid to basic research is in the fact that about ten U.S. companies account for almost 100 percent of U.S. research investment. These companies include IBM, AT&T, GE, DuPont, and Ford. Any surprises there? Not only is the overall amount of investment low, but the orientation of the research is often wrong. Generally, research is not innovative; instead, it is aimed at "perfecting" existing technologies. These are engineering refinements rather than innovations.

We've mentioned Merck on several occasions. In 1985, Merck appointed Roy Vagelos as its president and chief executive officer. He is a physician and biochemist who was originally recruited from a university campus to head their research area. He is seen as a bold and innovative leader, and he has said that one of Merck's basic beliefs is in "innovative research, that brings contemporary science to areas of human need."[3] Innovations can be done best in such an environment—where the innovative direction is clearly set right at the top.

If you look around at business successes and failures, with innovation as your yardstick, the performance of organizations takes on new meanings. Procter & Gamble beat out Lever Brothers by introducing Tide, a synthetic detergent. What was then National Cash Register took a financial bath and fired its CEO when its inventory of machines with electromechanical parts was rendered obsolete by the new electronics. The new financial products, such as combination cash-stock-credit accounts, are products of new technologies that were seized upon first by Merrill Lynch, then by others. Companies that have done well over the long term have been, by and large, successful innovators.

We've talked of successes and failures in dealing effectively with change. Many of today's organizations can still determine whether they will succeed or fail. Shouldn't this be of paramount concern to management, employees, shareholders, suppliers, and customers?

Innovation must permeate every facet of a company's operations. It is not strictly the province of the executive suite, nor is it solely a technological concern. In fact, it is quite often an asset of the organization's relationship with its customers, which should be based on the recognition that drills are only a way for customers to have holes. Xerox was able to invade the market of Addressograph-Multigraph with its new copier technology because it wisely focused on a new potential user—the secretary—rather than on the *competitor's* customers, the duplicating departments (which would, of course, become obsolete). That relationship with the customer, present or potential, must be constantly examined in light of not just the direct competition but of all influences affecting the relationship. Singer Sewing Machines faced years of declining sales, which it attributed to everything but the right reason: Singer's customers, women, were returning to the work force in droves, and their scarcer free time was now spent on fitness, sports, recreation, and so on. Singer didn't face product competition, it faced lifestyle competition. This same phenomenon was anticipated by the Simplicity Pattern Company. Simplicity—its very name is a wonderful marketing device—has positioned itself as the provider of sewing aids to the woman "on the go." Its techniques are easy to learn. It provides clear and concise written tips and is considering a set of instructional videotapes (marrying a growing technology with the demographics of its customers). Simplicity wants to be in the forefront of sewing services and products for the working woman. Singer, on the other hand, believed it was competing against Sears.

Innovation is occurring all around us. It is occurring because change is all around us. Innovation shouldn't be left to a few gifted individuals or saved for a few special instances. It should be part of management procedure and part of an individual's set of skills. We've tried to point out that innovative success is not the result of rare ideas or flashes of insight, but rather evolves from a methodical attempt to exploit change, to be master of your own fate.

Innovation is the tool of entrepreneurs, and virtually anyone can be entrepreneurial, within any type of organization. This simply requires a willingness to see change as opportunity instead of as threat and to employ some process for the orderly examination of change. Innovation is the entrepreneur's method of moving existing resources and assets from low yield and productivity to areas of high yield and productivity. Innovation needs to be organized. It is a process—no less a learnable skill than time management, planning, or delegating. As with any process, innovation's true value is that, once mastered, it is never forgotten, always available, and transferrable to others; it becomes the basis for objective judgments about complex events. The process of innovation affords the user the ability to determine prudent risks in terms of projected payoff and to identify implementation issues and strategic relevance. Indeed, history has proved that it is far riskier to attempt to innovate *without* a process than it is to consistently use one. The successful innovators discussed above had such processes. But the systems were not always formalized and often couldn't be articulated. This is fine for your grandmother's chicken soup (a little of this, a pinch of that) but not very effective for organizations. Hence the need to identify the steps and substeps in the innovative process and the requirement that they be discussed, practiced, and mastered. (Your grandmother actually did have a process. She either couldn't tell you, or wouldn't tell you, what it was.)

So, recognize the role of innovation, learn the process and skills of innovating, and *apply* these innovative abilities on a regular, disciplined basis. Don't wait for opportunity; look for it and exploit it. It is untrue that opportunity knocks but once—it is constantly knocking. The problem is that most people simply ignore it, or they open the door and don't recognize it. One objective of this book is to help you to hear that knock and to act on it long before most others.

Beyond One-Minute Solutions

Management is an area in which "quick fixes" are always sought and welcomed, and in the 1980's we've been overwhelmed by them. *In Search of Excellence* was an interesting examination of several successful organizations, but it was too readily embraced as the bible of management theology. What certain firms had done in their own environments with their own personalities does not translate easily into universal business practices. Hence, we witnessed the "embodied aphorism." For example, "management by wandering around" was a clever cliche—but you must know what you're looking for as you wander, how to do it (if even possible) in your culture without creating more problems than you're resolving, and what it means for your own productivity. "One-minute" approaches can be useful, but only when they are instituted in such a way that more than "one-minute results" are achieved.

The years ahead seem destined to be the age of entrepreneurial management. But how can this be said in light of what's passed before? Why isn't this just another quick fad to tempt managers to find the easy way out? For one thing, we know that management responsibilities will continue to grow more, not less, complex. The advent of technology in the workplace, for example, hasn't made management decision-making any easier. It's making all decisions tougher. The computer presents a high volume of data, but not knowledge in and of itself, and certainly not wisdom. Here are just some of the changes that will undoubtedly add complexity to the manager's job over the next ten years:

- Increased use of telecommunications
- Problems of security and confidentiality in data use and transmission
- The costs of new technology
- Changes in public perception (as has happened with health and smoking)
- The need to deliberately import foreign labor
- Virtually unpredictable economic and market volatility
- Increased foreign competition at all levels
- Fundamental demographic shifts
- Soaring new product introduction expenses
- Less upward mobility available within organizations

There are few forces in American business that can so successfully captivate the imaginations of others as the victory of an entrepreneur. How else can one explain the popularity with the "little guy" of a Lee Iacocca? Here's a multimillionaire, living a lifestyle that few can even visualize, turned into a hero of the common man because he had the wherewithal to buck the odds, take on the federal government and the competition, and turn a loser into a winner. Despite his organizational background and resources, Iacocca is seen as the embodiment of the American entrepreneur. That motivating force, if it can be harnessed and channeled, will be a prime mover of business and industry throughout the world over the remainder of this century. We are seeing an acknowledgment of this in François Mitterrand's "Silicon Riviera," in Scotland's "Silicon Glen," and in Margaret Thatcher's attempts to encourage entrepreneurship in Great Britain.

The innovator is nothing new in our society, but the *role* that the innovator plays and the ability to accommodate that role to the configurations of any size and type of business are new. What have been called "off–balance sheet assets"—human-resource energy, imagination, and focus—constitute a tremendous and powerful source of wealth. More and more organizations are learning how to exploit that asset.

SEARCHING FOR OPPORTUNITIES

Machiavelli wrote, "I have often reflected that the causes of the successes or failures of men depend upon their manner of suiting their conduct to the times." Nothing could be truer for management in these times of change and turbulence.

Searching for opportunities is aimed at being *pro*active. Its focus is on the future, where there are no "right" answers, only probabilities. All the facts aren't known—and can't be known—and managerial judgment is at a premium. The purpose of the innovation process is to organize management thinking in such a way that necessarily subjective judgments can be made using objective criteria. The results won't be known for some time. It's this innovative pursuit, however, that raises a company to a level beyond its competition and effectively carries it into the future, despite changing times. Problem solving is to treading water what innovation is to swimming for shore. We all must keep afloat, or people and organizations will drown. But unless we make headway, the chances are that something will come along and eat us, like sharks, or new technology—or tougher competition.

The tendency of managers to believe that they must take action often undermines innovative attempts. Fixing things is glamorous, universally appreciated, and full of short-term gratification. It's no surprise that we encourage entire cadres of problem-solving "junkies." There is an obvious need for effective problem solving and problem solvers, but *in addition to and in conjunction with* innovation and innovators. Both pursuits are as necessary as strategy and tactics. The key point to remember is that people can only use the tools they have at hand. If the only tool is a hammer, people will tend to look at every problem as if it's a nail. If all people know is problem solving, then "opportunities" will be viewed as problems to be fixed. This means that neither your maintenance nor your management will be able to keep the roof from falling on your head.

So *action* isn't the point, results are. And results come in a variety of flavors and sizes. Senior management must encourage (reward, train, exemplify, etc.) the accent on results and promote the role that innovation plays in achieving results. One of our clients was able to introduce innovation by the simple expedient of making certain people responsible for it. Management reasoned that certain people's positions were ideal for problem solving and others were better suited to innovating. Until both processes could be properly integrated in roles throughout the organization, it was easier—and more effective—to separate the responsibilities. When people speak of problems, they tend to look for cause (or worse, for blame); when they look for opportunities, they should be searching for change, which is usually on the horizon.

Hiding Poor Management

One could make a case that the emphasis on productivity, worker participation, and employee concessions in the recent past only partially reflects a concern about improving performance. The other concern is about covering the effects of poor management. Braniff originally filed for Chapter 11 protection not because of low productivity but because, in the rapture following deregulation, Braniff executives overextended their route structure. No amount of employee actions—not even 24-hour shifts by the mechanics or nonstop flying by the pilots—could have compensated for this error. And there is no other way to assess it. When workers try the wrong tool, or middle managers make guesses instead of reasoned decisions, or salespeople don't adequately prepare for the next sales call, we don't call it "unavoidable errors in judgment." We rightfully call it poor performance. Why should the top-paid people be exempt from such scrutiny?

There is a long list of companies that met financial Armageddon through executive ineptness, to which could be added those organizations that have not performed as well as they might have, owing to executive error, oversight, and sloth. People Express ultimately failed because its top officer, Donald Burr, couldn't adapt his system to the growth he voraciously pursued. Frank Borman hurt Eastern for years through his inability to reach harmonious relations with his unions. (Of course, the boards of directors are equally to blame, since these very same executives were forgiven their trespasses and, in many cases, richly rewarded for poor performance, while everyone else suffered.) Even firms that retain their economic vitality suffer badly from executive

Figure 1–1. Problem Solving Versus Innovation.

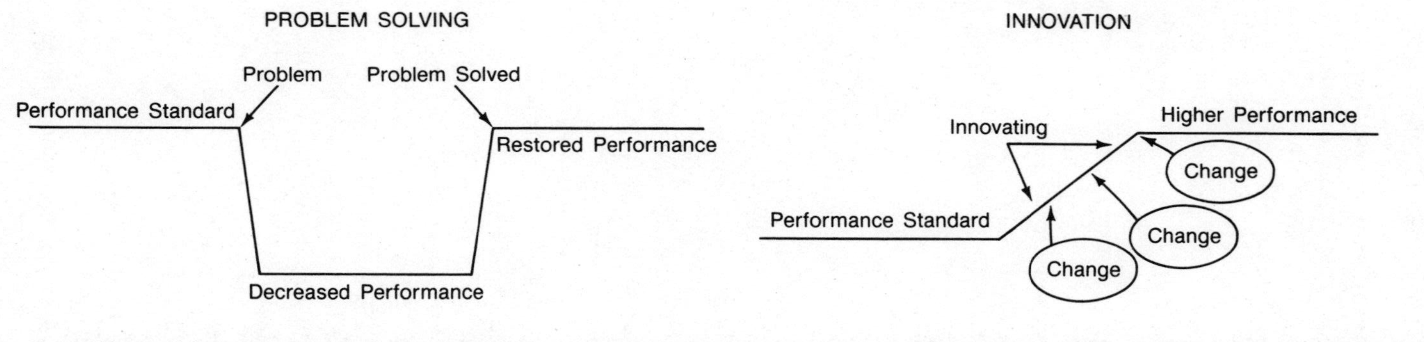

In *problem solving,* you can only be as good as you used to be.

In *innovating,* you can achieve new levels of performance, return, etc.

egos running wild. The dance macabre that Allied, Bendix, and Martin Marietta performed a few years ago hurt performance, shareholders, employees, and customers, all to sate a high-level ego battle.

We've found that a successful antidote to poor management—and to the general uncertainties of changing competition—is focused, deliberate, and systematic innovation. Too many companies pride themselves on the problem solving discussed above. It's simply not enough, and often the wrong starting place anyway.

STEPS TOWARD INNOVATION

Innovation will determine a company's ability to thrive, and perhaps survive, in the decade ahead. The following chapters will deal with the innovation process—what it is, how it works, and when to use it. We can and should expect competence among our managers, but we have no right to expect prophecy from them. So how is management to deal with new and usually unprecedented events? By employing the appropriate set of management skills and tools. Just as there are valid approaches to financial planning, delegation, decision making, and marketing, there are valid methods of innovation. We've distilled these methods from organizations and individuals around the world; they are based on common logic and are not industrially or culturally bound.

The unique approach we have developed to organize and systematically examine the innovation process, for any individual and any organization, is based on a practical search for, and prudent assessment of, opportunities.

We have separated the process of innovation into four basic steps:

1. Where do opportunities originate? *Opportunity search* is the organized examination of change to extract potential opportunities. It removes the "hit-or-miss" danger and obviates reliance on the rare "flash of insight" or "breakthrough idea." This step comprises ten areas of systematic search, which will be examined in Chapter 3.
2. The purpose of *opportunity assessment,* which is described in detail in Chapter 4, is to separate the high-potential opportunities from the others by providing pragmatic, basic criteria to evaluate each opportunity's potential.
3. *Opportunity development* is the refinement of your choice of opportunities through risk identification and containment. It means

that you are getting serious about the opportunities that you wish to develop further. We'll describe this step in Chapter 5.

4. *Opportunity pursuit* is the beginning of the implementation process. In this step, described in Chapter 6, you bridge the gap between conceptualization and implementation. It asks you to develop specific actions that will help to ensure the factors you need for success.

These steps are sequential, detailed, orderly, and logical; their purpose is to apply objective criteria to what might otherwise be subjective judgments.

In brief, our research has helped us design a process that is applicable to virtually any individual in any type of business, no matter what type of learning and understanding that individual employs to his or her best ends. The process is impartial and unbiased. While it can be used—as can any process—to promote parochial and biased points of view, such activities tend to become vividly visible when done within a system, where they can be examined by others.

Innovation isn't necessarily daring, bold action. But it *is* the willingness to look at things with an open mind and to examine change in an objective, confident manner. Some people, and a few organizations, can do this naturally. But it is not a very natural ability, since a great deal of our upbringing, schooling, and acculturation mitigates against it. Consequently, there is a need for a learnable, repeatable process to help us take such a view. There is no crime in not being naturally innovative, just as there is no crime in not being a natural swimmer. But if you know you will spend time in the water, it *is* irresponsible—and potentially fatal—not to learn to swim. And if you intend to earn your money in an organization, it is no less irresponsible and potentially fatal not to learn to be innovative. On the following pages we'll examine how to accomplish this.

But first, we'd like to shed some light on the most common misconceptions about innovators and the innovation process.

NOTES

1. Henry Ford, with Samuel Crowther, *My Life and Work* (Garden City, N.Y.: Doubleday, 1922).

2. Peter F. Drucker, *The Age of Discontinuity* (New York: Harper & Row, 1968); Richard N. Foster, *Innovation: The Attacker's Advantage* (New York: Summit Books, 1986).

3. "A Study of Organizational Values Held by Merck Middle Management," research project for Merck & Company, conducted by the Summit Consulting Group, Inc., East Greenwich, R.I., 1986.

Chapter 2
MISCONCEPTIONS ABOUT INNOVATION

Entrepreneurialism is not a topic that generates a confusion of definitions and responses when you ask people what an entrepreneur does or who an entrepreneur is. Unfortunately, questions about entrepreneurialism generate consistent responses, of which most are myths or misconceptions. A major reason for people not being more entrepreneurial—in their businesses and in their personal lives—is that they carry around an attitude about the entrepreneur that actually frustrates them from trying to be innovative, creative, or even a little different in their thinking.

MISCONCEPTIONS ABOUND

We strongly disagree with those who maintain that exploiting opportunities is a chaotic exercise. We believe that there is *always* a pattern at work in the systematic exploitation of change; however, it is too often unrecognized or disregarded.

Since so many of us have been inculcated with the idea that innovating is risky, we tend to steer clear of it. But innovation is not risky to the degree we've been led to believe. In fact, *not* innovating can be a far riskier proposition. Although misconceptions abound about who the entrepreneur is and what the entrepreneur does, we've isolated eight misconceptions that are especially inhibiting to those who might otherwise be innovative.

Misconception Number 1: Entrepreneurs are high-risk–takers. Most people believe that entrepreneurs are riverboat gamblers, constantly trying to draw to that inside straight. In fact, when we surveyed people, the most common response from the public was that an entrepreneur is a risk-taker. While entrepreneurs are indeed risk-takers, they are *prudent* risk-takers. They constantly calculate the amount of risk they are willing to take in return for the potential benefit of their innovation. We've asked a number of CEOs who have built very successful and fast-growing, mid-size companies—many of which are outperforming larger, multinational competitors—whether they had taken a lot of risks to build their organizations. We always received the same response: "I do not take any risk that I have not carefully thought through. I will never bet away the company on one roll of the dice."

There are some people who take excessive risk and sometimes have it pay off. These are the people who make good headlines and about whom we often read or hear. We shouldn't use the exception, however, to prove the rule. The fact is, most successful innovators carefully evaluate and analyze risk. *Every* decision contains risk, whether it's innovative or not. And some managers—in fact, some entire organizations—habitually choose to accept more risk than others. This is a personal predisposition, or part of an organization's culture. For example, the advertising industry will traditionally accept more risk in return for more benefit than would, say, the banking industry, which usually accepts only minimal risk in return for less benefit.

Venture capitalists receive a very high return on their investment when the company they've chosen to invest in is successful. That's because venture capitalists back very high-risk businesses. These businesses are often nothing more than a speculative gamble. Therefore, in return for putting money into such risks, a high return is justified. But that's not how innovation generally takes place. It's not a matter of venture capitalists backing a long shot, or of somebody betting the house and the family's future on one turn of the roulette wheel. Innovation is rather a question of careful analysis, methodical planning, and keeping an eye on what could go wrong.

The operating word for such risk-taking is "prudent." What is a prudent risk-taker? It is someone who is consciously aware of the risks, who evaluates those risks in terms of the benefit that may be realized, and who *then takes measures to try to reduce, minimize, and/or eliminate the risks.* Virtually no decision, especially an innovative one, is risk-free. Risk, however, can be controlled; it can be made tolerable, and it can even be mitigated. The successful entrepreneurs we've observed are very careful—very prudent—in their risk-taking.

Donald Trump, everyone's bold entrepreneur, is the epitome of the prudent risk-taker. "People think I'm a gambler," he says in *Trump: The Art of the Deal.* "I've never gambled in my life. To me, a gambler is someone who plays slot machines. I prefer to own slot machines."[1]

Misconception Number 2: Entrepreneurs are business owners, not employees. It's easy to identify business owners who are innovative in creating or building their businesses. The cookie industry's Mrs. Fields and Famous Amos quickly come to mind. Ray Kroc of McDonald's makes everyone's list of innovators, as does Fred Smith of Federal Express. Lesser "heroes," however, seldom come to the attention of firms other than those for which they work. This is regrettable, because most innovation occurs among people who work for someone else.

Some quick examples: 3M's Post-It Notes were created by one of its chemists; a fireproof corrugated box, the first of its kind successfully put into production, was created by people working for Digital Equipment; the frequent-flyer programs so popular today with airlines were pioneered by American Airlines' marketers; even something that has become as commonplace as the restaurant salad bar was not invented by a food expert or a restaurant owner, but by some now anonymous and unsung employee who suggested the idea. In 1956, a lab employee of 3M spilled a new chemical on her tennis shoe. A short time later, she realized that that area was not becoming as dirty as the rest of her shoe. What emerged from this unexpected event was a successful product called Scotchguard Fabric Protector. These innovations are the result of innovative individuals operating as employees, not owners.

We will talk more later in the book about the environment that is necessary to foster innovation and how to foster it within virtually any organization. For now, however, we will stress the point that while business owners can certainly be innovative, and often are, it is usually the employees who provide most of the innovative ideas. There are two simple reasons for this. The first is that there are simply *more* employees than there are owners. If two heads are better than one, several thousand heads are certainly better than one. So the fact that there are so many employees adds to the probability that innovative ideas will flow from this source. Second, employees are usually closest to the areas where innovation can take place. They deal on a daily basis with the product or service of the firm, with the firm's suppliers and vendors, with other parts of the organization, and, most importantly, with the *customer.* The employee is where "the rubber meets the road." A great deal of the success of quality circles and similar programs within organizations resides in the fact that employees are asked for their ideas, and those ideas often turn out to be highly innovative. Innovation, after all, doesn't deal only with a brand-new product or service. It can deal with an *improvement* in the existing scheme of things—a sales process, a manufacturing process, raising morale, creating a better image, or providing a new solution to an old problem.

Not long ago, we visited a furniture factory where management was constantly asking employees for new ideas. This was done on both a formal and informal basis, but it was done regularly. In turn, customers were invited to speak periodically with employees, and the customers were asked for their new ideas. This kind of cycle, we feel, is very important for effective innovation.

It is interesting to us that during the course of our research we found a great deal of money being invested in employee training. Yet virtually no money was being invested in providing employees with the skills necessary to improve and sharpen their ability to innovate. This is ironic, because the best source of innovation *is* employees, and employees have consistently demonstrated, when given the opportunity, that they *can* be and *enjoy* being innovative.

Misconception Number 3: Innovation takes place only in small firms, not in large firms. We are regaled with stories about small innovative firms, which are often in high-tech areas such as biomedical research, artificial intelligence, and telecommunications. And many large innovative firms began as small innovative firms: Apple, Domino's Pizza, and the growing airport shuttle services. But we could add other kinds of firms to this list. For example, the retailer Marks and Spencer in England has long led the way in innovation in an industry not famous for it. The *Wall Street Journal* has been highly innovative in establishing a national newspaper that reaches every subscriber and newsstand every morning, with reliability and accuracy. Mercedes-Benz, Michelin, and Sony are continually innovative. So innovation is not a function of a small firm, nor of a large firm—it can be a characteristic of *any* firm. (And notice its international scope as well. Innovative organizations are not culturally bound. They do not reside within any one group of industries or inside any one geographic border. This fact strengthens our conclusion that innovation is a discipline, a discrete set of skills, not a personality trait or a cultural trait. Creativity has no geography. More about this later.)

One question we had in mind was whether it was *easier* to innovate in small firms. And the answer, unfortunately for those of you who like black-and-white responses, is yes and no. If a large firm is bureaucratic, if its culture is antithetical to innovation, and if change is usually seen as threat, not opportunity, then it will be very difficult to innovate there. We found, however, that these same traits can exist just as easily in small firms. In fact, a small business owner with few employees can be just as bureaucratic, just as opposed to change, and just as threatened by change as any large company. Conversely, we found large organizations—Hewlett-Packard would be a good example—where innovation and the acceptance of change as opportunity is a part of the culture, and where bureaucracy is not a part of anyone's language. So not only does innovation not occur more in one size firm than another, it is neither easier nor harder in a small firm than in a large firm. (We want to emphasize that innovation is not simply "company culture." Changing the culture, without a process for innovation, is futile. Beware of those who view innovation as only a cultural issue. This applies to both the public and private domains.)

Misconception Number 4: Entrepreneurs have global ideas only and don't start with anything less than that. Early in our careers, we worked for a particularly aggressive marketing director. One morning, during an informal discussion in the hall, a subordinate rushed up to him with a brand-new idea. The idea clearly made sense, was highly innovative, and filled us all with a sense of enthusiasm and support. As he saw the ground swell of support rising, the animated subordinate concluded his presentation by saying, "I think this is worldwide in its impact." The marketing director never hesitated. He immediately replied, "No, it isn't. It's bigger than that."

Innovators aim at market leadership. Innovation is not meant to provide "me-too" or "also-ran" status. Innovation is meant to *lead.* Market leadership, however, can be gradually attained. While there are some ideas that are so good that they can immediately overwhelm the competition and vault one into first place, such ideas are relatively few and far between and should not be used as a model for implementing innovative ideas. The innovators we have studied generally start small. They are focused and precise in what they want to accomplish. In so doing, they are able to marshal their resources, direct their energies, and enlist the proper support to see their idea through to implementation.

An example of this is the personal diary called Filofax now in fairly widespread use. Filofax was a little-known company in London producing personal diaries that were unique in their flexibility. With a ringed binder and categories for diary, projects, notes, finances, etc., the diary provided a versatile and flexible tool for people who were dissatisfied with inflexible diaries, which could not be altered to their needs. In other words, at relatively low cost, the Filofax system enabled each user to custom design a personal diary. Filofax had a small but loyal following, although it was relatively unknown outside of England.

It was unknown, that is, until a visiting American businessman happened to find the diary, took a liking to it, and ultimately decided

innovative. In the latter case, you're taking existing resources and redeploying them.

Too often we have found managers thinking about going into new areas who base their decisions upon whether or not they can generate the *additional* resources needed to do so. There is insufficient attention given to the possibility or redeploying *existing* resources. The question isn't whether the existing resources are earning a return or not—the question is whether the return can be intensified and improved by investment in the new area. Generating a return isn't sufficient; generating the maximum return for a particular investment of resources—be those resources money, people, equipment, reputation, or whatever—is what innovation is all about.

Misconception Number 7: Innovation is brought on by flashes of genius. One of the great geniuses, Thomas Alva Edison, equated invention with perspiration. What we found is that, while there is the occasional kiss on the cheek by the muse, innovation is almost always the result of systematic, hard mental work.

The successes of companies with a history of consistent innovation—whether in the manufacturing or the service sector, whether 3M or Federal Express, Fuji Film or *USA Today*—belie the fact that people can be innovative only when struck with a bolt of creative lightning. Innovation will not necessarily flow, however, from simply *thinking* about innovation. The need is for a systematic, disciplined, pragmatic *process* that allows people to be innovative as a part of their daily work routines. Kroc was not struck from the blue with a thought about a fast-food hamburger chain. He learned of the McDonald brothers' restaurant when he was a milkshake machine salesman who received a large order from them. He had the curiosity to investigate their need for so many machines, the entrepreneurial sense to smell a good thing, and the innovative talent and ability to carefully evaluate and launch his enterprise. Fred Smith wrote a very disciplined and orderly college thesis on the hub-and-spoke method of delivering small parcels (which earned him a C−, no credit to his professor). Smith had thought long and hard about the weaknesses inherent in postal service and UPS delivery methodologies. True entrepreneurs don't wait for the flash of genius. They buckle down and get on with the hard work.

Flashes of genius and inspiration are, of course, always welcome. But if we sit around waiting for them in order to be innovative, the chances are our wait will be very long and lonely, while our competitors pass us by in the fast lane. The successful innovators we observed were systematic in their approach to innovation, no less so than they would be in their approach to solving problems, planning, or delegating responsibilities. They were continually examining how they innovated, ways to improve their process, and ways to transfer, transmit, and communicate that process to others. Still others innovated successfully without realizing how they did it. But once we worked "backwards" with them, observing their process and sequence of innovation, we and they both discovered that there was indeed a methodology at work. It is this methodology that we want to convey in this book.

Misconception Number 8: Entrepreneurs are born, not made. This last misconception is perhaps the most dangerous of all. That's because innovation is a *learnable skill.* And while some people may be born with the talents and abilities that will dispose them to become excellent innovators, most of us are not, just as we're not born with the natural abilities that will turn us into doctors or dentists, problem solvers or planners.

One of the factors we've observed that discourages people from trying to become more consciously innovative is the false belief that the ability to innovate is intrinsic. Not having "felt" innovative before, people feel that have little potential to become innovative. This is a terrible myth. We have found numerous examples of people learning a process of innovation and applying that process on the job, on a consistent and effective basis. Sometimes individuals are naturally able to grow to that level of competence. As they gather experience, education, and maturity, the right combination of factors synthesizes so that the individual suddenly finds himself or herself more innovative. They are not sure what to attribute this to, but they take advantage of it when they're smart enough to realize it. The trouble, of course, is that it is a long, laborious, and uncertain process. We found that people who have an open mind take the time to learn the skills that constitute a successful innovation process and have the mental discipline to apply that process on a regular basis. These people are just as innovative, if not more so, as those who have reached that condition "on their own."

The noted management authority, Warren Bennis, coauthored with Bert Nanus a book called *Leaders.*[2] In an effort somewhat analagous to our work on innovation, Bennis and Nanus interviewed a wide variety of leaders to find out what the traits of leadership are and whether leaders are made or born. Their conclusion is unequivocal: leaders are *made.* And while some people evolve into "natural" leaders, virtually anyone with certain basic skills and intelligence can learn the particular skills of leadership and can function successfully and effectively as a leader. The same is true of innovation. Innovation skills can be learned, practiced, communicated, and used individually or with others—with competency and purpose.

Some classic myths and misconceptions are generated and encouraged by the cliches with which we are raised. The prime offender among these is the one that goes, "Opportunity knocks but once."

In the first place, opportunity does *not* knock but once. Opportunity is knocking every day. The first part of the battle is to have in place the systematic process that forces one to focus on that fact. Opportunity is not a rare event, occurring with the frequency of a Halley's comet or a successful television situation comedy. Especially in this day and age of accelerating change and turbulence, opportunity is all around us.

If you were to draw a vertical line down a sheet of paper and place on the left the names of your close associates, business or family, who view change as threat, and on the right the names of those associates who view change as opportunity, we'd venture to say that the list on the left would be much longer than the list on the right. An interesting exercise is to then compare those on the left and right according to the amount of success you attribute to each of these people. In most cases, the most successful, no matter how you define success, will appear on the right-hand side of the line. Most people—including people we respect, love, and admire—are inculcated with the thought that change is threatening. Consequently, when change appears on the horizon, the first reaction is to circle the wagons or fill the sandbags.

TESTING: THREAT OR OPPORTUNITY

Even when we do hear the knock and open the door, we sometimes don't recognize what is staring us in the face. Are we looking at an opportunity? Or some dire threat? Or simply an unrelated event? Here's a quick test. Quickly jot down after each of these statements whether or not, in your experience, they would be a threat or an opportunity:

1. The company hires an individual from the outside for a newly created position that will be equal to your own.
2. The manager who has been your mentor for two years and has helped you rise successfully through the ranks is being transferred to another city.
3. For the first time, you discover that one of your key clients is having a meeting with one of your direct competitors.
4. The introduction of a new software package will probably render the software you're currently using obsolete.

5. A client who does a very small amount of business with you calls to complain about the poor service he has received recently.
6. A key client meeting is scheduled to last into the evening hours on your wedding anniversary.
7. The industry you work in is cited by a consumers' magazine as one that often engages in unfair practices with potential customers.
8. Through no fault of your own, tampering with your product in several isolated instances has created very adverse press coverage and a poor image in the eyes of the public.
9. It becomes increasingly apparent to you that there are at least three subordinates actively vying for your job.
10. Though no one is exactly sure what will happen, it is apparent that the government is going to change several of the regulatory conditions that pertain to the international operations of several of your clients.
11. In your marketing research aimed at introducing a new product in a different location, you find, to your utter amazement, that no one has ever used a similar product of any kind in that area.
12. You find that a copyright of yours has been infringed; someone is apparently copying your work.
13. During a presentation, a member of the audience openly and rather hostilely questions some of the basic assumptions underlying your argument.
14. A recently discovered weakness in your sales reporting system shows that projections made for the next three months are based on shaky data that may not be reliable.
15. In the middle of an overloaded day, with several deadlines facing you, a valuable subordinate bursts into your office and announces that she cannot go on under the current conditions any longer. Either things must be changed or she'll be forced to resign.

How many of these did you identify as opportunities, and how many as threats? Actually, all of them are based on documented instances of opportunities that can be exploited. We'll discuss three of them to demonstrate what we mean.

As in the fifth situation, a sales manager in San Francisco received a complaint from a client doing less than $5,000 in business yearly. The complaint concerned individuals whom the client was sending to certain training functions run by the sales manager's people. The client's complaint was that the training was uneven and that the instructors were not of a uniformly high caliber. The sales manager interpreted a complaint—the client bothering to pick up the phone to voice disapproval—as a sign of *interest.* The sales manager said, "I'm terribly sorry about the difficulty. In order to respond appropriately to your concern, may I have just a half hour of your time? I'll be happy to stop in at your office whenever it's convenient, but I would like to get the facts straight because such a quality lapse truly does disturb me." The client could hardly refuse, the sales manager visited, and a long and happy relationship was begun. Ultimately, that sales manager closed a $100,000 piece of business. The client? Bank of America. Even something as mundane and seemingly unpleasant as a complaint can turn into an extraordinary opportunity *if your attitude is oriented toward looking at such things as potential opportunity, not just as another burden on your day.*

The eighth situation refers, of course, to the Tylenol scare. Johnson & Johnson could have handled that problem in any of a number of ways. The most common way to handle such an apparent disaster would be to "sandbag." This is what we saw happen in the United States when customers claimed that the Audi 5000 was accelerating on its own. It's what we saw with the manufacturers of Bic, against whom claims were filed alleging that the lighters spontaneously erupted into flames. It's what we saw with the Ford Pinto. And on and on and on. Even though the tampering was not its fault, Johnson & Johnson chose a responsible and immediate course of action and, in so doing, seized an opportunity to enhance the repute and credibility of their firm. Not only did Tylenol bounce back in its market share, but the stock of Johnson & Johnson went up as well.

The eleventh situation is a stereotypical case. It is best exemplified by the old story of two shoe salespeople who were sent to an undeveloped area, one representing Company A and traveling down the East Coast, one representing Company B and traveling down the West Coast. A week after arriving, Sales Rep A sent a telegram back to the company: "Please send fare to return home. No one here wears shoes." Sales Rep B, however, sent his telegram: "Send reinforcements immediately! No one here wears shoes!" Just as some people look at a glass and see it half filled while others see it half empty, people also tend to look at change and interpret it in different ways. Many of them see only the threat that change brings, even when only the slightest degree of threat is present. But a chosen few can see opportunity because they're ready to recognize it and *can* recognize it when it does indeed occur. This is simply a question of holding a different attitude, but it is the key aspect of recognizing opportunity. This is the major difference we have observed between a person who is "perceived" as an entrepreneur and one who is not. The entrepreneur sees change as a constant source of opportunity.

We hope you'll see that innovation is not a function of "culture change," a phrase that's come to mean everything and nothing. When people innovate, "cultures" change anyway. But trying to change to a more innovative culture without providing individuals with the skills and processes with which to innovate is like changing a ship from sail to motor power without installing engines. There are disciplined ways in which to do this, and we will begin exploring those steps in the next chapter.

NOTES

1. Donald Trump and Tony Schwartz, *The Art of the Deal* (New York: Random House, 1988).
2. Warren Bennis and Burt Nanus, *Leaders: The Strategies for Taking Charge* (New York: Harper & Row, 1985).

Chapter 3
OPPORTUNITY SEARCH: WHAT'S THAT KNOCKING AT THE DOOR?

"Every significant step in every field is taken by an individual who has freed himself from the way of thinking held by associates and friends who may be more intelligent, better educated and better disciplined . . . but who have not mastered the art of the fresh, clean look at old, old knowledge."[1] Relatively few companies take the time to think through entrepreneurialism thoroughly, much less encourage fresh thinking, as Edwin Land describes it above. That means that innovative efforts, when they do arise, don't receive the necessary support from the organization. According to a study conducted by Arthur Young and Company, the Institute for Innovation, and the Forsythe Group, nearly 75 percent of the respondents in a survey of 500 senior executives listed innovation as a priority, yet half of them said that their companies were unlikely to review the causes of innovative failure.[2] In other words, there was no systematic plan in place for learning from their mistakes. Further, two-thirds of the respondents reported a lack of *any* coherent innovative planning process within their organizations.

Recently, Minneapolis police announced that they would no longer assist the motorists who lock themselves out of their cars each year. In that city alone, the number of motorists who manage to do that is estimated at about 20,000 annually. Within three days of the announcement, local entrepreneurs had set up a 24-hour "doorman" service called Car-Help.[3]

In 1987 the Bureau of Business Practice, a Prentice-Hall subsidiary, introduced three programs that deal with smoking issues. The programs include videotapes, and they cover the rights of both organizations and individuals under the different no-smoking rules being promulgated at various government levels. Smokenders, Inc. cites about thirty calls a week from organizations seeking its help, and it estimates that New York City enrollment alone will soon triple as a result of that state's new smoking regulations, which took effect midway through 1987. A Massachusetts firm that offers smoke-ending programs using hypnosis reports that its corporate work has doubled in the past year.[4]

One of the latest and hottest items to strike the attention of affluent buyers is, of all things, the traditional fountain pen. It's been called a "power tool for the 80s."[5] According to the Writing Instrument Manufacturers' Association, sales of fountain pens are up 30 percent over the last year alone. Prices start at about $3 and go up into the stratosphere ($6,000 and more being not uncommon). The average fountain pen costs between $100 and $250, depending upon materials.

At this writing, more than 250 companies nationwide are conducting AIDS research, and that number is steadily growing. According to Mervyn Silverman, president of the American Foundation for AIDS Research, "People are waking up to the fact that AIDS is a big problem, that it's here to stay for a long time and—to be quite crass about it—there's money to be made."[6] It is estimated that approximately 1.5 million AIDS patients will require treatment over the next five years. The Center for AIDS Control in southern California projects that over that period, victims will spend $1.5 billion a year on drugs to treat the disease. Most companies in the field, however, are not concentrating on cures and treatments. They are searching for more reliable and faster diagnostic tests to detect AIDS in the bloodstream.

The latest estimate reveals that there are about 20 million people enrolled in various airline frequent-flyer programs in the United States alone, and that in 1986 those enrollees took free trips worth over $1 billion.[7] American Airlines introduced its Advantage Program, the first frequent-flyer program, in 1981. It was designed to be a short-term publicity move, and nothing more. The frequent-flyer idea has evolved, however, into the largest, and probably best, marketing device ever produced by the airlines. The programs have completely changed spending and marketing patterns and have become a primary technique employed by the airlines to create customer loyalty. Surveys show that, for business travelers, the programs rank second only to scheduling convenience in criteria for choosing airlines. Every day, the airlines—now joined by rental car organizations, hotels, and credit card companies—are devising new and different ways to lure people into their programs.

MINING CHANGE FOR THE ORE OF OPPORTUNITY

What do all of these events have in common? They reflect innovations that few had thought to pursue. They also represent the first step in an innovation process that successfully implements new ideas: *the search for changes that will produce opportunities.*

We talked earlier about opportunity knocking not only once but repeatedly, the problem being that most people either don't hear the knock or fail to recognize opportunity when they finally open the door. Opportunity search, however, goes beyond just listening for the knock. It is a proactive, assertive, even aggressive, step that involves systematically examining the environment for potential opportunities. We like to look at it as "mining" various sources of change for opportunity.

The problem with waiting for opportunity to knock, even if you do hear it and recognize it, is that opportunity may have knocked at other doors previously. The chances are that by the time opportunity comes to you, someone else has already begun to exploit it. You might be able to catch and pass them, or you might devise a better way of exploiting the opportunity, but in either case you're starting off at a disadvantage. It's sort of like knowing there's a bear in the woods and taking a seat outside of the woods to wait for the bear to come out. Of course, the bear may come out and approach you, but on the other hand, it may never come out. Or it might come out at some other place. If you go into the woods, the likelihood of your finding the bear and being the first one to find it is much greater. There's some risk involved in entering the woods, that's true, but the rewards would seem to far offset the risk.

We've found that even among successful innovators, there were many who weren't sure how and why they found the opportunities that they did. When we helped them to articulate and codify the approaches they were unconsciously using, they were able to increase both the frequency and the quality of opportunities that they discovered. And a key point to be considered throughout this book is this: while it is helpful to increase the ability of anyone to innovate, the biggest payoff is in helping those who are already fairly good at it. In baseball, the .200 hitter who is able to improve to .225 has made some significant strides. But the .310 hitter who is able to improve to .317 has made an even more valuable advance, both personally and for the team, if his extra hits are batting in more runs. So even those who are successful innovators by "gut feel" are able to significantly enhance their effectiveness by doing what they always do on a more methodical and uniform basis.

One of the ironies of the opportunity search step in the innovation process is that it doesn't deal with a dearth of changes, but rather a plethora. We found that innovation is stifled when people are faced with the multitude of changes that we discussed in Chapter 2, but have no effective method for sorting them out. One of the reasons we feel that change often appears as threat rather than as opportunity is that *the absence of a process* for examining change results in a very intimidating situation. Consequently, people are prone to throw up their hands and run from change because they are not quite sure how to deal with it. The greater the change, the worse the effect. As a result, even those people who don't view change as threat have no effective mechanism for intelligently identifying and selecting those changes that embody the greatest sources of opportunity.

The importance of having a process is that it enables us to sort through the multitude of changes facing all of us in the workplace every day (or for that matter, in our personal lives). By having a process, we can not only deal with change intelligently ourselves, but we can talk about it accurately with our colleagues. This response—changing a potential threat into a potential opportunity by having an effective means of understanding and dealing with it—can apply to many elements in our lives.

Whether it's public speaking, writing, learning to operate a piece of equipment, analyzing a balance sheet, learning to swim, or any other skill to be mastered, the application of a *process* is the key element. A process enables the individual to objectively face the task and conquer it, mitigating the emotion and fears that often get in the way of mastery. This phenomenon is especially true for those who seek to become more entrepreneurial.

THE TEN SEARCH AREAS

We have identified ten areas that can be mined in the first step of the innovation process, the opportunity search. You might have others to add to the list, and certainly the list is not exhaustive. But we have established that these are the ten areas that most good innovators use, either consciously or unconsciously, to mine for opportunities. Some of the areas are obvious, and some not so obvious. Some will seem especially applicable to you and/or your business, and some will seem somewhat irrelevant. Some of the areas are easy to explore, others more difficult. The point is, *all* of these areas need to be examined on a consistent and systematic basis if you are to be truly innovative.

As we go through the list and provide examples, we think you'll begin to see what we mean by some people not hearing the knock of opportunity and others opening the door but not recognizing it. Not all airlines embraced the frequent-flyer programs at an early stage. In fact, some were dragged kicking and screaming into the programs only recently, despite earlier protestations that they would never provide such giveaways. Not all fountain pen manufacturers were in a position to capitalize on the instrument's newfound cachet. The difference is not "being in the right place at the right time," and it's not luck. The difference is that some individuals and organizations are constantly on the lookout for opportunity. Being on such a lookout means having the wherewithal to examine the following areas of potential opportunity with diligence and regularity.

These are the ten sources of opportunity that we have found to be the most universal. That is, they represent those areas that most companies are able to examine most of the time. Your particular circumstances and your particular organization might differ somewhat. But don't be too quick to modify, add to, or delete from the list. Our research overwhelmingly supports the conclusion that these areas are truly universal, and that organizations are sometimes lax in examining them.

Opportunity Search Number 1: Unexpected Successes

Most people accept success readily enough, and most recognize it when they experience it. Relatively few individuals, however, make that key determination that allows them to build still further on this success. After all, if failures can be exacerbated, why can't successes be exploited? Unexpected successes can happen to both your own organization and those of your competitors. Most people explain away unexpected successes as temporary aberrations that will soon disappear.

Ray Kroc wondered why he was selling so many milkshake machines to a small restaurant run by the McDonald brothers. Only through examining the cause of an unexpected success, as Kroc did, are we able to genuinely exploit change and create innovative new approaches.

Another example occurs in the personal computer field. When personal computers were developed and launched, their manufacturers naturally hoped that they would be successful, but it's safe to say that no one ever dreamed that they would be as successful as they have been. After all, when Steve Jobs and Steve Wozniak were tinkering in their garage—building what turned out to be the first Apple computer—no bank would underwrite their fledgling company, and the two entrepreneurs themselves saw their market as much more limited than it ultimately turned out to be. The personal computer's success was striking enough, but the scope and degree of its success were truly shocking and, consequently, most unexpected.

Jobs was able to exploit that success by very rapidly building one of the major computer firms in the world, Apple Computer. Others, however, were able to capitalize on his success as well. From giant companies like IBM and Tandy to brash newcomers like Osborne and Franklin, other innovators who identified that unexpected success were also able to exploit it very early in the marketing cycle of personal computers. Those who were smart enough to see the market for peripherals and support services also profited nicely. Perhaps the best-known example is Bill Gates, who founded and continues to run Microsoft. Gates has recently become one of the few billionaires in the industry, joining the likes of such legends as Bill Packard and David Hewlett, and he is believed to be the *only* person to make a billion dollars in the software field.

The magazine industry was also able to take advantage of the unexpected success of personal computers. Scores of them cropped up; some specialized in particular models of computers—such as *Rainbow,* published for the Tandy color computer users—and other magazines appealed to the programmer and/or "hacker." (*Rainbow,* originally a four-page flier produced on a copy machine, is especially interesting. In the space of just a few years, it's grown into a magazine of well over one hundred pages, with a sizable subscription base that has enabled its publisher, Lenny Falk, to diversify into computer conferences, other magazines, and other publishing activities.)

Note that in the magazine and peripheral equipment fields, there might have been many more losers than winners. The difference we've noted between the losers and the winners—especially when they entered the market at the same time—is in the better planning of winners. (This will be discussed further in subsequent chapters, when we deal with the development and pursuit steps.) It's not good enough to be first in the market, you also have to be *best* in the market. Our point, however, is that unexpected success can lead to early entry into the marketplace if individuals are adept enough to recognize it and adroit enough to quickly take action on it.

Here are some questions that we have found useful to help mine opportunity from unexpected successes. By asking yourself these questions, and by formalizing them as part of your work routine, you will tend not to overlook unexpected successes, whether yours or someone else's. Not all of the questions may apply, but we find that by asking them in a disciplined fashion, answers to those that do apply tend to reveal the raw material you need to be innovative.

- What unexpected *product* success have you had recently?
- In which *geographic* areas have you had unexpected success recently?
- In which *market/industry* segments have you experienced unexpected success recently?
- What *customer* segments have provided unexpected success recently?
- What unexpected successes have your *suppliers* had recently?
- What unexpected successes have your *competitors* had recently?
- Which of your *technologies* has had unexpected success recently?
- What unexpected *customer/user* groups have bought from you recently?
- What unexpected sources have asked to *sample/distribute/represent* your product recently?

We find that in asking these questions, you will generate ideas and opportunities that you hadn't thought of before. Again, you might wish to add questions to the list that pertain to your particular industry, culture, or position. But it's not the order of the questions or the particular questions that generate opportunity; it's the *discipline* of asking questions similar to these as a regular part of your work routine.

Once the questions have provided you with some raw material to work with, the next question to ask is, "What specific opportunities and ideas can be developed from these unexpected successes?" Note that the process is not completed simply by naming the events that led to unexpected successes. On the contrary, that's only the beginning of the process. The next step is to seek opportunities that result from these unexpected successes. Some of the successes may not present opportunities, at least, not as far as you can see. (This is one of the reasons why it's always helpful to use this process with other people. It provides a systematic framework within which a combination of ideas and perspectives creates a dramatic result.)

The third step in the process of looking at unexpected success for opportunity is to develop *specific* concepts for new *product/service/customer/markets or improvement* ideas to capitalize on that opportunity. This third step is crucial for two reasons. First, we established earlier that innovation isn't complete until it's implemented. Ambiguous or vague or even global ideas can't be implemented. Second, the suc-

Figure 3–1. Unexpected Successes.

UNEXPECTED SUCCESSES	NEW OPPORTUNITIES	NEW {PRODUCT/SERVICE, CUSTOMER/MARKET, IMPROVEMENT} IDEAS
INTERNAL		
EXTERNAL		
Requests from small independent distributors to market our products	Franchise network to reach new customers	Franchise local wholesale distributors in San Diego and Los Angeles as test program

© 1987 Decision Processes International. All rights reserved.

ceeding steps of the process depend upon those specifics for the purposes of assessment and development. It's one thing to say, "Let's expand our marketplace." That's a "motherhood" kind of idea; it's hard to argue with, but also very hard to do anything about. It's quite another thing to say, "Let's expand our market into three New England states," or, "Let's expand our market into small retail stores located in cities of under 50,000 population in New England." The more specific the idea, the easier it is to evaluate, and the safer it is to try to implement. This is one insight that separated the winners from the losers in the peripheral computer equipment field.

So whether it's specific new customer/market segments, or specific new geographic markets, or specific new improvement programs—or specific anything else—it's important that the flow appear as it does in Figure 3–1. For example, if you asked yourself which unexpected sources have asked to sample or distribute your product recently, and you found that although you have a captive field force you've received requests from small, independent distributors, you've now got a piece of information—a small piece of "ore" from your mining process—that you can attempt to exploit. In the next part of the search process, you'll ask, "What opportunities and ideas can be developed from this piece of information?"

Let's say that the idea is to develop a franchise network of distributors who can sell your product to smaller businesses that your sales force doesn't have the time, inclination, or financial support to reach. Note that we've gone from an unexpected success—a new source asking to distribute your product—to an opportunity: a possible franchise network. But that opportunity is still very broad. It's hard to evaluate the worth of a national franchise operation, much less to develop a plan to implement it from scratch. So this brings us to the third part of this step: what *specific* new customer or market could you seek?

For the purposes of our example, let's say that you want to reach the artists' supply stores with an annual volume of less than $1 million, operating in suburban shopping malls. Consequently, you might develop the specific idea of franchising local wholesale distributors of art supplies who currently have such stores as their customers. You might further choose to begin on a pilot basis in a fairly densely populated area. So the idea might evolve into launching the pilot solely in the suburbs of San Diego and Los Angeles.

The important thing to understand is that you've moved from an unexpected success—new requests to distribute your products—to a very specific idea: franchising certain wholesalers to represent you in a clear demographic area that you are currently not reaching. Whether this idea is a "good" one will be determined later as you assess it and develop it. For now, however, you've been able to take an event that otherwise might have escaped your notice and have turned it into a potential opportunity. This is the kind of management activity that should be going on *daily*. Although at first glance it may seem like you don't have the time to do it, the process greatly reduces time demands in the long term. More importantly, however, you can't afford *not* to do it.

Opportunity Search Number 2: Unexpected Failures

If unexpected success is our first area, and one that's easily overlooked in the search process, our second area is its converse: unexpected failure. Unexpected failure may be failure on our own part or failure on our competitor's part. In either case, we're looking for failures that we didn't anticipate.

You might say that *any* failure is one that wasn't anticipated, and that everyone plans to be successful in everything they do. To a certain extent, that's correct. But we're talking here of rather dramatic unexpected failures. That is, failures that occurred when great success had been anticipated.

The Edsel's failure was a signal to at least one Ford executive that people were no longer buying cars simply because of the nameplate or because they wanted no more than basic transportation. People were beginning to buy cars because they wanted to make a lifestyle statement. The individual who realized this was named Iacocca; he subsequently introduced the single most successful new car in the marketplace up to that point: the Mustang. The Mustang was the first, real lifestyle automobile. Ford could have simply licked its wounds and walked away from the Edsel debacle with its tail between its legs. Instead, through Iacocca it learned a lesson that it used to create an innovative new car. And in turn, this unexpected success on Ford's part generated the Camaro for General Motors. It seemed to take General Motors only about half a day to produce the Camaro after the Mustang's success. (Unexpected successes or failures—and virtually all of the other areas discussed below—should be applied to the competition as much as to your own organization.) Unexpected failures occur on a fairly regular basis. We are not trained, however, to look at them in terms of opportunity.

As mergers and acquisitions have accelerated, we've seen many firms—even apparently large, healthy, and vigorous firms—unexpectedly fail to fight off acquisition attempts or succeed in doing so only by severely damaging their worth, using "greenmail" or "poison pill" remedies. These unexpected failures of even large organizations to defend themselves against the raiders have presented opportunity to an entire industry of financial analysts, bankers, and attorneys who now specialize in helping organizations to fight off such threats. The unexpected failure of organizations to be able to do this themselves has created opportunity for someone else.

Failures are disliked, so people tend to defend the failure (protect themselves) rather than try to find out what caused it to happen and how to turn it into an opportunity.

Here are some questions we have found useful to help discover unexpected failure.

- What unexpected *product* or *service* failures have you had recently?
- In which *geographic* areas have you had unexpected failure recently?
- In which *market/industry* segments have you experienced unexpected failure recently?
- What *customer* segments have proved to be unexpected failures recently?
- What unexpected failures have your *suppliers* had recently?
- What unexpected failures have your *competitors* had recently?
- Which of your *technologies* has had unexpected failure recently?

Figure 3–2. Unexpected Failures.

UNEXPECTED FAILURES	NEW OPPORTUNITIES	NEW { PRODUCT/SERVICE CUSTOMER/MARKET IMPROVEMENT } IDEAS
INTERNAL		
Poor rating in consumer magazine evaluation of ballpoint pens	Change pen's image, to differentiate from competitors'	Redesigned, modern-looking "new" pen. Advertised as "the pen for modern managers"
EXTERNAL		

© 1987 Decision Processes International. All rights reserved.

- What *traditional customer/user* groups have unexpectedly failed to buy from you recently?
- What *traditional market* segments have unexpectedly declined recently?
- What *traditional distributors/representatives* have unexpectedly not met quotas recently?

Again, you might wish to add, delete, or modify items on this list. As with unexpected success, the first step is followed by two others. Once you've identified an unexpected failure in any of these areas, you should then ask, "What opportunities and ideas can I develop from these unexpected failures?" Having developed the opportunities, you should then go to the third step: "What specific new product/service ideas can I think of to take advantage of this opportunity? What specific new customer/market segments could we serve? What specific new geographic markets could we seek? What specific new improvement programs could we develop?" (See Figure 3–2).

For example, if you're in the business of manufacturing writing implements, and you find that your ballpoint pens have received from a consumers' magazine a poorer quality rating than your competition, this is certainly a cause for concern. It's also, quite clearly, an unexpected failure. You expect your ballpoint pens to stand up well against the competition, and you've tested them and manufactured them with that expectation. But something has gone wrong.

What opportunities can you develop from this unexpected failure? Well, one such opportunity might be to change the entire image of the pen. At the moment, it's fairly nondescript, and it also competes against several competitors' pens that look very similar. Since it seems that some reworking is going to be necessary anyway to improve the quality problem, perhaps this is an opportunity to change the image and make the implement much more distinctive and appealing to the consumer. So the specific new product idea that might come from this opportunity is to redesign and modernize the pen so that it can be launched with a new advertising campaign that legitimately calls it a *new* writing implement. This may seem like a natural evolution to many of you, but in all candor, most managers would regard this as simply a problem to be solved. Even worse, many would regard it as an excuse for finding *blame.* Finding who is responsible for the quality lapse might be helpful—and certainly, finding what materials or parts of the assembly process have resulted in the poor quality is very important—but all of these are problem-solving efforts. They can only restore the pen to where it once was: just one of many similar products. Your opportunity here is not just to fix the problem but to improve upon the situation vastly by developing a new, more competitive, writing instrument.

Many of you may be saying, "Okay, but that's easy to say and hard to do. After all, you're talking about substantially increasing expenditures by launching a new product rather than just fixing the old one."

Maybe so, maybe not. Those kinds of issues will be taken care of in the opportunity assessment step, which will take into consideration cost and benefit as well as other factors. But it is important to remember our earlier definition: entrepreneurs *redeploy* resources and assets to increase yield and productivity. There is certainly nothing wrong with asking the question, "Can we increase our yield and our productivity, using our current resources, by launching a new implement rather than merely fixing the old one?" That question can be answered first on paper and can be intelligently discussed and further examined as the opportunity gets more specific. *But the question has to be asked in the first place.* On too many occasions, managers are simply rushing to make the "fix" and are not looking at anything further. This is the difference between problem solving and innovation, and the difference between being an also-ran and a market leader. This is why the innovation process must be part of a manager's daily routine. When failure crops up, it's not good enough merely to fix it. One must ask, "What opportunities may await us here?" At any meeting, there should always be a person in the room who looks at failure and is able to say, "How can we benefit from this? How can we profit from this? What opportunities might this present for us?"

This may seem like an odd thing to do at a time of failure, but that's exactly what's being done by innovative individuals and innovative organizations all over the world.

Opportunity Search Number 3: Unexpected Events

Unexpected events, like successes and failures, can be internal or external. But here, external events needn't be happening only with your competitors. Such events go on in the world at large.

Unexpected events are ideal sources for true innovative thinkers. For example, with the onset of aircraft hijacking in the 1960s, entire new industries were born. While many organizations undoubtedly considered hijacking an event totally unrelated to their basic business, some firms that happened to have the technology to produce metal-detecting equipment realized that a tremendous opportunity lay before them. Those that recognized it first—were "lightest on their feet," so to speak—and redirected their resources toward this emerging need entered a multimillion-dollar industry rapidly and gained a predominant market share. This is a classic example of redeploying existing resources and assets for better yield and productivity. Demonstrating that this is not a process that deals only with tangible goods and manufacturing, some firms supplying guard and security services similarly realized that such services would be needed at airports, and they quickly moved into that vacuum. Even though in many cases this was merely a matter of redirecting some training efforts and pursuing contracts from a different source, not all such firms were able to recognize the opportunity and, more importantly, to rapidly act on it. This example epitomizes what we mean by recognizing opportunity knocking and being able to do something about it.

We've also mentioned the unexpected tampering with over-the-counter medical products, such as Tylenol; those firms intelligent enough—innovative enough—to redirect their efforts toward a new need responded with tamper-resistant packaging. The AIDS epidemic has provided tremendous opportunity for the manufacturers of condoms, the more innovative of whom rapidly sought out advertising and distribution methods that never would have been considered earlier. Consequently, we are seeing condom advertising on television

Figure 3–3. Unexpected Events.

UNEXPECTED EVENTS	NEW OPPORTUNITIES	NEW { PRODUCT/SERVICE CUSTOMER/MARKET IMPROVEMENT } IDEAS
INTERNAL		
EXTERNAL		
Resort city wins bid for trade association convention	Expand taxi business to include services catering to conventioneers	Place ad promoting personal tours that are unique, with advance discount available

© 1987 Decision Processes International. All rights reserved.

for the first time, but only from those manufacturers fast enough, wise enough, and bold enough to pursue this avenue.

- What unexpected *external* events have occurred recently?
- What unexpected *internal* events have occurred recently?
- Have any *expected* external and internal events *combined* in an unexpected way recently?

Having identified these unexpected events, you should now ask, "What opportunities and ideas can be developed from these unexpected events? What *specific* new product/service/improvement ideas can be developed within this opportunity?"

Here's a simple example, but one that shows the process can be applied to any kind of operation and in a variety of ways. Suppose you run a taxi service, and you operate a dozen cabs in a town of about 25,000 people. You learn that a resort city ten miles away has landed a major trade association convention. You discover that the convention will last for eight days, covering one complete weekend. You learn from your hotel contacts that at least 600 rooms have been booked for the convention, and that most of these bookings include spouses.

Most taxi companies would probably feel good about this unexpected new business as a "shot in the arm." Perhaps some might even arrange to lease more cars and acquire temporary drivers to increase their presence at the airport and the hotel at the beginning and end of the conference. Your cab company, however, is going to be even more innovative than that. You're going to take out an ad costing several hundred dollars in the trade association's advance mailing and catalogue. What will the ad say? It will say that your company offers individualized tours of the entire resort area and its coastal communities, by the half day or full day, at reasonable prices. The ad will go on to say how unique it is to take this tour with a personal driver, as opposed to being part of a large group. And, of course, if arrangements are made prior to the convention, there will be a significant discount on the cost. You can schedule the tours between 10:00 a.m. and 4:00 p.m. so that your lucrative morning and evening airport business won't be disturbed. You'll be seeing all of your cabs utilized for the entire day, however, if the idea takes off.

There would certainly be some scheduling problems: taking care of regular customers, selecting which drivers are capable of providing such a tour commentary, and scores of other details like these. The subsequent assessment step, however, will determine whether or not such investments and redirections are justified by the increased business. For now, it's an idea that ought to be considered, at least on paper. Does this seem farfetched? Could a small cab firm develop this kind of innovation? Could it make these kinds of adjustments?

It's not farfetched at all. We've seen ideas like it work dozens of times. Unexpected events, obviously, occur every day all around us. Most are seemingly unrelated to our jobs, our lives, and our interests. We hope we've been able to point out, however, that this isn't necessarily the case. Only by a disciplined approach, using some simple questions, can you really determine whether an outside event presents opportunity for you, no matter how divorced from your life and your job it might initially seem.

Opportunity Search Number 4: Process Weakness

Process weakness is that weak link, or even missing link, in one of the systems, procedures, or processes of a business. This does not have to be a continuous flow process, such as a papermaking process or a chemical manufacturing process. It can be a financial reporting process, a sales forecasting system, a personnel evaluation system, a distribution process, or any of a host of other things.

The process weakness in the post office's failure to provide a guaranteed system of reliable next-day delivery created the phenomenal success story of Federal Express. (Of course, long before Frederick Smith entered the scene, United Parcel Service was already capitalizing on the post office's weakness in distributing parcels and packages.) Process weakness should be investigated as thoroughly in one's own organization as in that of the competition. Sometimes merely correcting a process weakness isn't sufficient. True innovation relies upon replacing the weakness with a dramatic improvement.

It should be noted that Federal Express has also had its share of entrepreneurial failures. For example, Zap Mail proved to be a very expensive failure, and Fred Smith was forced to abandon it. This does not mean that the innovative spirit and abilities of Federal Express are waning. On the contrary, it simply shows that the company continues to be innovative, even after its initial, dramatic growth. Despite the best-laid plans and the best of systems, not all innovations will be successful. But we know that the only way to have successful innovation is to try to innovate on a regular basis. Since only *prudent* risk is accepted by successful entrepreneurs, Federal Express survived the failure of Zap Mail. We know of few examples as good as that of Federal Express to prove these points.

One of the best examples of process weakness we know of is that of the Internal Revenue Service. This is a system and an operation that infuriates taxpayers, while inefficiently collecting revenues for the government. We've often speculated that if the collection process were turned over to private enterprise, or if a company could compete with the IRS to collect government revenues in return for some portion of the total billings, not only would a private company thrive but the government would collect far greater revenues at much less cost. Perhaps this is an innovation we should hope does *not* occur.

One of our newspaper clients recently implemented an interesting innovation regarding one of its process weaknesses. It had found that its system of attracting and collecting billings on classified advertising was highly inefficient. Time deadlines sometimes prevented advertisers from placing last-minute ads. Moreover, telephone lines stayed busy, operators made mistakes, typographical errors occurred for which the newspaper had to "make good," and a host of other gremlins constantly plagued the system. The newspaper took a look at this weakness and decided that, rather than improve the training of the operators, extend phone hours, or double-check copy, it would take the opportunity to enhance classified advertising in general by giving the advertiser a role in the process.

The newspaper developed computer software that enabled advertisers to place classified ads directly in the paper by using their own personal computers or business computers and a telephone modem.

In this manner, advertisers could place ads right up until the last minute before the deadline, only hours before the printing of the newspaper. There was no longer a need for a "middle man," since the advertiser was communicating directly with the newspaper's computer. An even greater advantage was that any typographical errors were the fault of the advertiser, and therefore the newspaper was no longer responsible for them. Advertisers were highly in favor of the scheme because it gave them total control of their advertising and much more flexibility in their use of the newspaper's space. Consequently, this weak link in the process of the newspaper was not improved but totally revamped to provide a greater opportunity and a greater return on *existing* assets and resources. What began as process weakness became what one might consider an unexpected success, and the newspaper is exploiting this by looking into software that will enable display advertisers to have similar flexibility, including the ability to design their own layout.

- What *self-contained* processes exist in the organization?
- What *weakness* or "*missing link*" prevents better process performance?
- Why do some processes perform *better at some times than at others?*
- What *bottlenecks* do each of these processes have?
- What process weaknesses among our *competitors* might we be able to improve on?

Let's suppose that you're the manager in charge of billing at an electric utility for a large city. The weakness in your process is the billing and return procedure: Because the address is written illegibly, many of the customers' payments are delayed, misplaced, or misaccounted. You've identified a way to remedy this: by providing self-addressed envelopes. You've already calculated that the cost of the self-addressed envelopes will be more than made up for by the fact that the increased efficiency in obtaining payments will generate a great deal more in interest on deposits made earlier at the bank. The self-addressed envelopes will also cut down on your expense in following up payments that aren't made or have been lost. Being an innovator, however, there is more you can do than just correct the situation.

Since your billing and receipt procedure is a process weakness, what opportunities might this provide for you? Since the bills have to go out monthly anyway, and the customer has to open a bill, why not use this as an opportunity to communicate other things to the customer besides the amount they owe the utility? What *specific* ideas might result?

Figure 3–4. Process Weakness.

SELF-CONTAINED PROCESS	"WEAK" OR "MISSING" LINK	NEW IMPROVEMENT OPPORTUNITIES
Customer billings and collections	Inaccurate/delayed/ misdirected/lost return payments	Preprinted return envelopes and inserts promoting other products/ services and customer education issues

© 1987 Decision Processes International. All rights reserved.

One idea is to provide a brief flier or brochure in the billing envelope, detailing extra products and services the utility provides: wiring inspections within the home, energy-saving evaluations, "smart" energy-saving appliances, or a host of other items that the consumer might not be aware of. By providing this kind of information in the billing envelope, along with a self-addressed envelope, you're not only fixing a current problem but you're being innovative: you are seeking to increase your yield with little more than a minor change in your billing procedure.

Think about the bills you're getting from Exxon, Texaco, Diners Club, or a host of other organizations not usually associated with providing consumer products. You'll find that you're being offered everything from television sets to telephones, VCRs, insurance, and snow tires. Companies have enlisted their bill distribution methods in the effort to enhance sales.

Opportunity Search Number 5: Changes in Industry and/or Market Structure

Industry and market structure changes, such as those we see all around us in the telephone, cable television, and health care industries, and in the internationalization of many businesses, are excellent sources of opportunities. These changes have given rise to the likes of MCI, Sprint, and other alternative telephone carriers, as well as to cable television, satellite dishes, and peripheral businesses such as decoders. It's interesting to note that at least one AT&T competitor, Sprint, has chosen not to do battle with AT&T against its traditional strengths. Instead, Sprint emphasizes the higher quality of transmission that its total optical fiber provides over AT&T's older technology and copper cables.

In a similar manner, Humana is restructuring traditional approaches to health care by providing mass-produced services that achieve economies of scale and by making profit-oriented assessments of facilities use. Similarly, "elder care" and "catastrophic care" are viewed by health care providers as markets to be exploited.

- What major changes are occurring among your *customers?*
- What major changes are occurring in your *geographic* markets?
- What major changes are occurring within your *market/industry* structure or in the *conduct of your business?*
- What major changes are occurring among your *competitors?*
- What major changes are occurring within your *regulatory* environment?
- What major changes are occurring in your *supplier* relationships?
- What major changes are occurring in how you *finance* operations?

Let's suppose that you're a catalogue publisher, specializing in consumer items for the home. Over the past several years, your growth has come from the fact that two-income families spend less discretionary time visiting retail outlets. Consequently, you've concentrated your list on an "upscale" lifestyle and have emphasized the ease and availability of ordering things through the mail. Lately, however, you've noticed that many in your market have begun to move back to the cities. This gentrification has made it much easier to shop at retail outlets, which are now near the home. It's no longer necessary to drive to malls or to suffer through relatively limited selections in the suburban boutiques. By walking from a brownstone down the block, your customer can now find a variety of stores and outlets catering to almost every need. In view of this market structure change, you can make a highly innovative move.

That move is to open your own retail outlets in the cities—not to compete with other retail stores but to represent your catalogue and the tremendous name recognition it's developed with its audience. This move gives you an increased access to a market already familiar with the quality of your goods, services, and good name. By this time, many of you will have recognized exactly what catalogue operators

Figure 3–5. Changes in Industry and/or Market Structure.

INDUSTRY/MARKET STRUCTURE CHANGES	NEW OPPORTUNITIES	NEW {PRODUCT/SERVICE, CUSTOMER/MARKET, IMPROVEMENT} IDEAS
Gentrification: customers move back to the city	Let customers come to you, instead of you approaching customers only through catalogues	Open retail outlets in fashionable areas of New York and San Francisco, keeping catalogue name on front of stores

© 1987 Decision Processes International. All rights reserved.

like The Sharper Image have indeed done. They've opened trendy, attractive stores in New York's newly refurbished South Street Seaport area and in similar locations in Boston and San Francisco. It's an interesting example of an innovator applying old ideas in new ways.

Opportunity comes quite readily from industry and market structure changes if one is able to use a process to see through the turbulence and the often frenzied surroundings.

Opportunity Search Number 6: High-Growth Business Areas

High-growth areas are areas, products, services, markets, etc., inside or outside the organization, that are growing faster than the gross national product or faster than the general population. Because organizations need growth to perpetuate themselves, areas of high growth are always attention-getters. Such a development occurred immediately after World War II, when the baby boom created new demand for certain products and services, demand that was insufficiently anticipated by most organizations. More recently, it's occurred in the phenomenal growth of the personal computer and peripheral computer equipment industries.

Many years ago, the retailer R. H. Macy noticed increased sales of small appliances, sales far beyond expectations. Macy's management had previously decided that small appliances should play only a limited role in the overall offerings of such a department store. In spite of brisk sales, they decided not to increase supply but simply to sell out what supplies they had and to keep their small-appliance sales an established percentage of overall sales. In so doing, of course, they completely missed the opportunity presented by a society in which "modern conveniences" had become equated with success and in which anything that could reduce housework was being eagerly sought out. This void was more than filled by other retailers and other sources of such appliances.

You and your colleagues should be constantly asking, "What parts of our business are growing faster than economic or population growth?" Yet we find that high growth is one of the most overlooked of the ten areas of opportunity search. One reason is that people are more attuned to looking for *poor* growth than to looking for *high* growth.

- What *parts of your business* are growing faster than economic or population growth?
- What *other businesses* are growing faster than economic or population growth?
- What potentially high-growth businesses related to yours are *dominated by only one or two companies?*
- What parts of your *competitors'* businesses are growing faster than economic or population growth?
- What parts of your *customers' or suppliers'* businesses are growing faster than economic or population growth?

Home videocassette recorders, as everyone suspects, comprise a fairly high-growth area, with penetration beyond most original estimates. What isn't generally known, however, and what has been aggressively pursued by the more innovative retailers, is that the two-VCR home is becoming increasingly common. Just as having two televisions and two telephones is no longer rare—and in many cases is the norm—so, too, is having two VCRs in any given family. One can be viewed while another is time-shifting programs, or one can be used by the parents and another by the kids. In any case, here's a high-growth area on top of a high-growth area, if you will. This is the kind of nuance that true innovators look for in their pursuit of opportunity.

Figure 3–6. High-Growth Business Areas.

HIGH-GROWTH BUSINESS AREAS	CHANGES OCCURRING	NEW {PRODUCT/SERVICE, CUSTOMER/MARKET, IMPROVEMENT} IDEAS
INTERNAL		
VCRs	Two-VCR households are becoming accepted (not luxury, but necessity, especially for multi-TV homes with children)	Promote "two-for-one" sales with playback-only unit provided with purchases of top-of-the-line, full-feature model
EXTERNAL		

© 1987 Decision Processes International. All rights reserved.

Opportunity Search Number 7: Converging Technologies

Here we are speaking about two or more technologies that perhaps singly do not represent opportunity, but when taken together represent substantial opportunity for those willing to look for it. For example, the marriage of the computer to the videodisk has created an interactive learning combination that several training firms have been in the forefront of perfecting. This combination allows the learner to see a vignette presented on a television monitor and to respond to it on a keyboard. Various branching operations allow the learner to view different vignettes depicting the outcomes that the learner has chosen.

A variation of this can be seen in automobile showrooms, where the customer may ask the salesperson a question such as, "What does the car look like in red?" If the salesperson doesn't have a red car available, there's a major problem. What is a salesperson to say to this customer—"Go visit our competitor who has a red one, and if you like it, come back here and purchase it from us"? The salesperson knows that if the customer leaves to find a red car, he or she probably won't be back. And the various color swatches and chips of paint that were provided can't provide a very good idea, even for the most conceptual and visual among us.

Figure 3–7. Converging Technologies.

CONVERGING TECHNOLOGIES	NEW OPPORTUNITIES	NEW {PRODUCT/SERVICE, CUSTOMER/MARKET, IMPROVEMENT} IDEAS
INTERNAL		
EXTERNAL		
Cellular telephone and electronic computer "mail" and messages	Improve field sales productivity with enhanced communications	Provide cellular phone in each salesperson's car, with link to electronic computer message center in home office

© 1987 Decision Processes International. All rights reserved.

Now, however, the customer can sit down at a computer/videodisk machine and, in effect, design a car. If the customer wants to see a Thunderbird in fire-engine red, one is shown on the screen (under the best possible lighting and environmental conditions, of course). If the customer wants to see the sports package, or the leather upholstery, or how a tire is changed, all of this can be viewed while sitting in one place under the watchful eye and with the attentive help of the salesperson.

- What technologies in your business are *converging* or *merging?*
- Which of your technologies is now being joined to *outside technologies?*
- Which of your technologies can be more effective if *deliberately* converged?
- What would be the *ideal* convergence of technologies in your business?

We recently saw a client combine the computer with cellular telephones. Previously, the client's salespeople had to make sure they called in several times a day to check on messages from their clients, prospects, and management. This required finding a phone, sometimes in awkward situations, and often with little or no privacy.

Thinking about his existing computer system, our client looked at the new cellular phone operation available in major metropolitan areas and came up with a truly innovative approach. All messages for salespeople are now stored in the computer's "electronic mailbox." The computer can be accessed from cellular phones installed in each salesperson's automobile. Consequently, the salespeople can always call in complete privacy. Moreover, they can use otherwise nonproductive time spent driving from one destination to another to check in with the office. Salespeople can also get directions to new sites and help when they're lost, all without leaving the car and losing time. The salespeople feel less dependent on their lunch hours and on secretaries, who had the inevitable problem of reading handwritten messages that were occasionally illegible. Hence, the convergence of these two separate technologies in an innovative manner has dramatically increased productivity.

Opportunity Search Number 8: Demographic Changes

Demographic change is interesting because sometimes it's subtle, and sometimes it is rather rapid. Demographics, of course, can involve such things as the age or education of your users, their income, where they are living, the mix of users, and so on and so forth. We mentioned gentrification earlier. We have found that many managers are still laboring under the illusion that the youth market—the so-called Yuppie market—is the preeminent buying market in the economy. This is not true. Rather, the "gray market" is expanding most rapidly and possesses the greatest potential buying power. This is why we see innovative companies providing distinct products and services to this user group. Things like "leisure communities" for people over a certain age have long been with us, but now we are seeing leisure resorts, travel agencies catering only to people over fifty years of age, and advertising aimed specifically at this age group.

How dramatic is demographic change? Here's a quick example from the United States: A man who is married to his first wife, who is the sole breadwinner in his family, and who has two children and a house in the suburbs, now represents less than 4 percent of the population. That's right, less than 4 percent. Not too many years ago, such a man would have seemed relatively "average." Today, however, with nearly 50 percent of marriages ending in divorce, almost 60 percent of women working outside the home, and more people employed by McDonald's than by U.S. Steel, demographic change can sneak up on us. Sneaky or not, however, it presents tremendous opportunity.

Figure 3–8. Demographic Changes.

DEMOGRAPHIC CHANGES	NEW OPPORTUNITIES	NEW PRODUCT/ SERVICE IDEAS
USER AGE		
USER EDUCATION		
USER INCOME Widespread retiree affluence resulting from IRA accumulations, beginning to be withdrawn in 1999	Provide for "absorption" of these funds as mandatory withdrawals take place	Provide new product—"Grandchildren Love Trusts"—for those wishing to establish college funds for preteen grandchildren, with minimal tax burden
USER MIX		

© 1987 Decision Processes International. All rights reserved.

- How is the *age* distribution of your customers and users changing?
- How will the *educational* level of your customers and users change in the next few years?
- How will the *income* distribution of your customers and users change in the next few years?
- How will the *geographic* distribution of your customers and users change in the years ahead?
- How might the *buying habits* of your customers and users change in the years ahead?
- What are the *customer demographics* that might change over the years ahead?
- How will the *mix* of your users and customers change in the next few years?

Demographic changes have long been used by direct-mail advertisers, political analysts, and aggressive marketers. Such changes, however, can represent opportunity for *anyone,* irrespective of his or her business and its products or services. (See Figure 3–8.)

Sometime in the years ahead, there will be an unprecedented infusion of money into the economy. Where will it come from? It will come from the IRA funds that will have built up over a 10- or 20-year—or even longer—period and will have provided a retired and educated population with a greater affluence than has ever been known before. The tax legislation that provided for these IRAs will be producing a dramatic monetary change in the population, decades after its enactment. In the next few years, the first large amounts of IRA money will start to be withdrawn and will be followed by larger and larger withdrawals. The opportunities will abound for people who are capable of helping IRA recipients to invest, spend, safeguard, and provide for the ultimate redistribution of those funds. Remember, you heard it here first.

Opportunity Search Number 9: Changes in Perception

This search area is unique in that it is the only one that can be consciously manipulated. After all, that's what advertising is all about—attempting to achieve changes in perception. Note that changes in perception are not changes in the facts themselves, but rather changes in the way people choose to *interpret* the facts.

Such changes in perception are common and dramatic. For example, people have been driving while drinking since the automobile was invented nearly a century ago. But it's only within the last few years that the true danger of driving while under the influence of alcohol has been underscored and brought to the public attention, with the resulting stigma on such activity. Similarly, perceptions of personal fitness, the hazards of smoking, the importance of preventing certain diseases, and other health matters have changed radically.

In working with a client near the top of the Fortune 500, we noted one division experiencing increasing difficulty in selling microfilm for archival use. Ten years ago, microfilm was perceived as high-tech. Today, electronics is perceived as high-tech, and microfilm as "low-tech." Although microfilm is still the best available alternative (the product and its usefulness have not changed), in most cases the customer's *perception* of the product has changed substantially.

How have entrepreneurs capitalized on perception change? Well, we've seen the decline of the "happy hour" and the singles bar and the corresponding rise of the health club. It seems as though the place to meet young, eligible, single people these days is no longer in the singles bar but in the health club. (And, as an editorial aside, it seems to us that many of those people in the health clubs have never worked up a healthy sweat.) Tobacco companies have rushed to diversify their businesses in response to the public's growing concern over the perils of tobacco smoking. Brewers and distillers have moved toward lighter alcoholic beverages, low calorie beverages, and nonalcoholic beverages. From the same supplier, you can now buy sodas with any combination of no-caffeine, no-sugar, and no-carbohydrates.

- What changes are occurring in how your products and services are *perceived?*
- What changes are occurring in the *values* of your customers?
- What changes are occurring in the *lifestyle, image, and status* of your customers?
- How will changes in perception *affect* your customers and suppliers?
- For what *new purposes* have customers purchased your products and services recently?
- What *intangible* reasons are customers developing to support your products and services?
- What societal, peer, and normative *pressures* will affect your products and services in the future?

While some changes in perception come belatedly, such as recognizing the risks of tobacco, and some merely achieve greater prominence, such as the danger of driving while drinking, others are deliberately created. For example, perfume advertisements create the perception that its use will create erotic interludes. Such advertising is not limited to the private sector. The armed forces have been able to completely change the public perception of the military in the post-Vietnam era. The Marines now advertise for "a few good men." The Army advertises its ability to train people for high-paying professional careers after their service career is over. The Army's current advertising slogan is, "Be all that you can be."

Too often, companies waste money trying to improve the product, or improve its distribution, or better train the sales force—when in fact what's needed is a change in the public's *perception* of the product. No matter what Schaeffer or Waterman appears to be able to do, neither has been able to create the same perception of quality as has Cross Pens. Rolex has dominated the public's perception as the best watch on the market, as has Mercedes as the high-class automobile. These last two examples clearly point out one benefit of achieving a changed perception: you can then *charge* for the perception. Let's face it, no Mercedes is worth the amount of money that it costs to purchase one. The engineering may be superb, and the styling may be excellent, but a good portion of the purchase price represents the perceived prestige that rolls along with it. Years ago, some forgotten innovator determined that the public will pay for perception, and it's a lesson that Mercedes et al. have not forgotten.

So perception is very important in innovation. You should examine your current products and services, both for their vulnerability to

Figure 3–9. Changes in Perception.

YOUR CURRENT PRODUCT/SERVICE	CHANGES IN PERCEPTION	NEW {PRODUCT/SERVICE, CUSTOMER/MARKET, IMPROVEMENT} IDEAS
Bodybuilding apparatus for schools, gyms, and other institutions	Women can and should engage in bodybuilding and other rigorous fitness efforts	Produce "designer line" for women's exercise equipment, with weights, colors, design, and endorsement appealing to women who want to engage in bodybuilding exercises while maintaining feminine image

© 1987 Decision Processes International. All rights reserved.

perception change and for your ability to change perception of them. Perception change is fickle—it can happen on the spur of the moment, almost before you realize what's taken place. Only by diligently and consistently scanning the environment and surveying your own operation as well as those of your competitors will you be able to stay ahead—and therefore profit from the opportunity—of perception change.

Opportunity Search Number 10: New Knowledge

New knowledge probably represents the basis for most new patents granted, year in and year out. But very few of those patents ever result in a viable, marketable product.

New knowledge has some intrinsic barriers. For one thing, new knowledge, or pure invention, often takes large amounts of research and experimentation, as well as a long time to develop. Secondly, the window of opportunity in new knowledge is often severely limited. For example, new knowledge is usually applicable for only a brief time. Or if it is applicable over a long period, new knowledge is difficult to protect from the inroads of competitors, patents and copyrights notwithstanding. So new knowledge is perhaps the most

Figure 3–10. New Knowledge.

NEW KNOWLEDGE IN YOUR BUSINESS	IMPACT OF THIS NEW KNOWLEDGE	NEW {PRODUCT/SERVICE CUSTOMER/MARKET IMPROVEMENT} IDEAS
Ink that will write and not smear underwater	Provides for fine precision writing in intermittent or continuously wet conditions	Form joint venture relationship to provide such pens with all tropical fish supplies so that professional hobbyists can keep records in aquariums, hatcheries, etc.

© 1987 Decision Processes International. All rights reserved.

difficult of all the areas of opportunity search, yet it nonetheless produces significant opportunity. We've consistently mentioned Merck & Company as an example of one of the more innovative firms. Over the years, Merck's innovation has come almost exclusively from its research and development efforts—from the new knowledge it's been able to generate through its laboratories and scientists. New knowledge can be an excellent source of opportunity if it's explored carefully and systematically.

- What new knowledge has recently become known about *your business?*
- What *combinations* of knowledge have created new insights into your business?
- What new sources of *information* about your business have recently been tapped?
- What new *patents* or *discoveries* have been announced relating to your business?

The laser is an excellent example of new knowledge. It has provided industrial, military, communications, and health care uses. Note the diversity of application that this one bit of new knowledge has created. Don't assume that new knowledge has to be specifically related to your products and services. The true innovator finds distinct and discrete *applications* of new knowledge that can benefit his or her particular business. Superconductors are the latest new knowledge breakthrough, with major implications for industries ranging from communications to transportation.

We have discussed the areas of opportunity search in the order in which you are likely to find success in them; that is, unexpected success consistently leads to more successful innovative efforts than does new knowledge. Yet we have seen many companies concentrating their efforts in areas at the bottom of our list. They bet their future on new knowledge, looking for the "miracle" opportunity, while ignoring the more probable opportunities found at the top of our search list. Nevertheless, all of these sources—and perhaps more—need to be systematically mined by every organization wishing to ensure the best possible chances of success in stimulating its innovation efforts.

These are the primary sources of innovation, the areas from which opportunities can best be derived. But what of the bright idea, the miracle cure, the conceptual breakthrough? They have all played their part, to be sure, and patents are issued every day on inventions based on them, some of which are successfully implemented in the marketplace. Yet they aren't the source that entrepreneurs generally rely on, nor are they the sources that lend themselves to a process. They are not repeatable on demand, cannot be taught and learned, and are seldom immediately applicable. Great ideas shouldn't be ignored when encountered; but neither should they be relied upon to assist in innovation on a daily basis. It's like finding a 10-dollar bill in the street. You're glad it was there, and you certainly can use it, but you wouldn't leave your house hoping to find one to pay for the groceries that day.

By systematically examining these areas as part of your daily business routine, we can almost guarantee that your organization's ability to innovate will benefit dramatically. When looking at each of the search areas, it's important to first *identify* the particular event—unexpected success, demographic change, change in perception—that might be of relevance. This is the ore that you've found in the ground. The next step is to look for opportunities in those events—that is, to extract the ore from the ground. The final part of the search step is to establish *specific* and detailed ideas that can flow from this opportunity. This is the cleaning and polishing of the ore that we hope you've discovered to be gold.

The important effort is to identify opportunities and not to worry about their source. It is not important whether an opportunity came from unexpected success, a competitor's process weakness, or a high-growth area. Just *identify* it. The process does not give points for neatness. In other words, it doesn't matter which area generates the opportunity, so long as it *is* generated. That's why we have posed so many different process questions. The goal is to identify the opportunity, no matter what its source.

Of course, the next step is to visit the assayer's office to determine how much the gold is worth and to make sure that it's not simply fool's gold. We call that the opportunity assessment step, the subject of the next chapter.

NOTES

1. Edwin Land, "The Intuitive Manager," *Bottom Line,* 30 March 1987.
2. John Naisbitt and Patricia Whittington, "Helping Companies Hatch Offspring," *Success,* May 1987.
3. *Wall Street Journal,* 14 April 1987.
4. *Wall Street Journal,* 16 April 1987.
5. Troy Segal, "The Fountain Pen: A 'Power Tool' for the '80s," *Business Week,* 22 September 1986.
6. Deborah Denaru, "Lucrative Market Developing in AIDS Research," *Providence Business News,* 19 January 1987.
7. *New York Times,* 19 April 1987.

Chapter 4
OPPORTUNITY ASSESSMENT: THE GOOD, THE BAD, AND THE UGLY

An effective process of innovation enables you to assess opportunities in terms of risk, benefit, cost, ease of implementation, and relation to business strategy, among other factors. As stated earlier, the problem is not that opportunity knocks but once, but that it is knocking every day in the multitude of changes facing organizations. We have found—initially to our own surprise, and always to the surprise of clients—that organizations have a surfeit of opportunities. Once they are recognized, the real challenge of management is to differentiate among them and to select those that promise the greatest potential benefit within the company's basic business plan.

Just as in problem solving and decision making (where difficulty usually arises not from an absence of information but from too much information, which is often contradictory and conflicting), problems in innovation arise from having too many opportunities to deal with and too many factors to consider. Therefore, there is all the more need for some filtering or winnowing process that will allow a manager to make intelligent choices about the correct opportunities to pursue.

By systematically following up with the assessment step, you are able to complete the search step with an eye toward mining areas for opportunity, not worrying at that point about the risk, threat, and peril. As in any good idea-generating process, the search step should produce new ideas without negative comment either from yourself or from colleagues. Its intent is to provide a clear-thinking and freethinking base from which to generate new ideas. It's the assessment step that provides the means to evaluate these ideas and to critically analyze them for their ability to meet entrepreneurial parameters as we have defined them: shifting existing assets and resources from areas of low productivity and yield to areas of high productivity and yield. Consequently, you are also able to evaluate existing uses of those assets and resources in this step.

Among other concerns, the assessment step asks you to gather and evaluate information in response to such questions as:

- What is the cost of implementation?
- What is the cost of sustained operation?
- What is the benefit in dollars, morale, quality, service, market share?
- How difficult will implementation be?
- How related is this idea to our strategy, our business, and our direction?

Assessment begins the analysis of how well the opportunity fits your business and your culture and helps to analyze risk, so that "pet projects" aren't insulated from real consequences. Finally, the assessment step begins the identification of critical factors that will determine the ultimate success of the idea arising from the opportunity, so that when and if you reach the implementation stage you are armed with a plan to ensure success. Too many great opportunities die on the drawing board or are mortally wounded when the conceptualizer and the implementer are different people or different groups. Innovation has to result in successful implementation, and the assessment step is the first stop along that route.

The assessment step is to priority setting what a reverse somersault dive is to a belly flop into the pool. While both will get you into the water, the former is safer, more pleasing, more accurate, and much more likely to generate applause.

How are new ideas and alternatives currently decided upon in your organization? Often the loudest voice or highest position will determine which approaches receive proper resources. Sometimes it's the squeaky wheel. At other times, it's simply the safest—or at least, what is *perceived* as the safest—course of action that's chosen. Still other times, what's been done in the past determines future actions—not terribly daring decision making, but at least it's terribly safe. The assessment step provides an objective template not only for judging new ideas against each other, but also for judging new ideas against existing uses of assets and resources. In this way, the true value of redeploying those assets and resources to achieve higher productivity and yield can be visibly and tangibly evaluated.

THE FOUR ASSESSMENT AREAS

We have isolated four key areas that we believe encompass the essential criteria for assessing new ideas. These four areas have subsets and can be further divided, tailored, and modified to best represent your organization and its culture. We have learned, however, that, as in most management practices, the ideal process is not necessarily the most complicated or complex one. Rather, the ideal process is one that can be used quickly by an individual and/or efficiently and harmoniously by a group, no matter now disparate its members. Consequently, while we've found that many groups tend to modify or detail the criteria underlying each major area, we have not found any group or organization that feels the need to change the four basic areas.

The first two areas are the traditional ones of cost and benefit. Anyone engaged in innovation should want to know fairly concisely and rapidly: What will be the cost of implementing this opportunity? And what will be the benefit of implementing this opportunity? Here are some of the generic questions usually asked:

Questions: *Cost*

- What top-caliber *resources* are available to support this opportunity?
- What are the *investment* requirements?
- What *track record* do others have in this area?
- Do we need to *hire* people?
- Do we need to secure *new expertise?*

Questions: *Benefit*

- Can a successful *pilot* translate into a successful enterprise?
- Does the *funding* weaken our financial position?
- To what degree is the *risk* worth the benefit?
- Specifically, how long until we see the expected *results?*

The following is a checklist of considerations to assess that contribute either to the cost or potential benefit of an opportunity. An evaluation of each will allow you to create a framework within which the cost-benefit relationship can be clarified and scrutinized.

Checklist

Cost	*Benefit*
People	Market share
Materials	Return/profit
Equipment	Prestige
Research	Service
Marketing	Earnings
Legal	"Fallout"
Promotion	Safety
Time	Quality
Pilot	Morale
Contingency	Growth

The other two major areas we've chosen are less obvious. But we've found that these areas, and the nuances they represent, often spell the difference between success and failure in the implementation of even excellent opportunities. These two areas are *strategic fit* and *difficulty of implementation.*

By strategic fit, we mean the degree to which an opportunity fits a company's direction. The opportunity need not fit exactly at first, provided that it promises to fit well in the near future. So the issue concerns not just today's strategy but also the direction that the company is pursuing.

When an opportunity is pursued irrespective of strategic thrust, the results are usually disappointing, or even damaging. Several years ago, Exxon decided to enter the office information business and began an operation known as QXT. Despite the infusion of massive amounts of money and the work of some highly talented people, QXT was a disaster, and Exxon eventually dissolved it. The office products market simply was not a part of Exxon's direction and strategy as a company. That is, it could not be *made* a part of the nature and direction—the fabric—of Exxon's business. Consequently, it was doomed to failure from the start.

Similarly, People Express seemed to be making good headway toward its own goals when it was forced out of the sky by its acquisition of Frontier Airlines, a traditional carrier that did not fit very well with the vision and direction established by Donald Burr for People Express itself. Innovation is not a question of unbridled enthusiasm spinning off into every direction of the compass at once. It is a question of organized, purposeful, and *focused* attempts to improve the organization's products, services, markets, and operations in general. The end result of innovation has to be an enhanced ability to meet and exceed the organization's *business goals.* Consequently, innovation must be undertaken within the purview of corporate strategy. This helps to set parameters and establish limits.

There are companies, however, whose nature and direction embrace such profit-driven strategy. For them, any purchase, and virtually any opportunity, is sound so long as that opportunity generates the appropriate degree of profit. At one time, for example, the Transamerica Company owned a chartered airline, a relocation service, Occidental Insurance, Budget Rent-a-Car, and the United Artists motion picture company, among other organizations. Transamerica's strategy was focused on profit: as long as a certain profit was generated, the subsidiary fit into the overall structure. If the profit couldn't be generated, then the subsidiary no longer fit. We've seen the same hold true for conglomerates such as Gulf & Western and the "old" ITT under Harold Geneen.

So strategic fit is extremely important, particularly when the opportunity is global and far-reaching in nature. Obviously, the opportunity to revamp the company's billing system, or to change the interview process for new candidates, may almost automatically be strategically acceptable. Even here, however, we can see some subtleties that can make all the difference. For example, in 1986 Kodak hired approximately ten electronic engineers for every one chemical engineer. This was virtually a direct reversal of what it had been doing a decade earlier. Even something as tactical and specific as the kinds of people recruited and hired should reflect the organization's strategy. The engineers Kodak is now hiring clearly demonstrate the direction that Kodak's research and development effort is taking; their selection also says a great deal about where the company expects to be deriving its revenue over the years ahead—electronically oriented products instead of chemically based products.

The relative ease or difficulty of implementation—the fourth and final area—really refers to the "organizational immune system" that every company seems to possess. We've seen numerous cases of good ideas that presented high benefit at reasonable cost, within the organization's strategic framework, but were rejected by this immune system.

A contemporary case in point is Honda's marketing of its new Accura automobiles. These cars represent Honda's entry into the luxury auto marketplace. Honda made a decision at the outset that these luxury cars could not be sold through the same dealerships that were handling the rest of the Honda line. It was apparently felt that the Honda "sales culture" would not be conducive to selling these high-end automobiles. So Honda's decision was to market these new cars only in separate dealerships, and even then, only in separate dealerships that maintained a certain physical distance from any nearby Honda dealerships. This decision was taken even at the risk of offending current Honda dealers. Honda clearly believes that its organizational immune system is not ready for the luxury auto.

Another example of implementation difficulty is in the training profession. The educational and training industry that caters to business is estimated by the American Society of Training and Development to be approximately a $30-billion business. Yet we estimate that no single training firm has even one-half of one percent of this market! One of the reasons for this is that training companies specialize in only a very few areas. For example, a company might be adept at marketing time-management techniques. Another might be skilled at selling sales and marketing training. Still another might be oriented toward strategy formulation. Whatever the specialty, we found that virtually none of these firms are able to sell a wide array of training products and services that cover the panoply of management skills.

The marketing and sales departments in these companies always fall back onto the path of least resistance and sell what they are *comfortable* selling. Consequently, although logic would seem to dictate otherwise, it's the norm to find a large organization like Prudential Insurance or Ford Motor Company or American Express dealing with ten or even twenty outside training vendors, each representing a different management skill or discipline. When training firms have attempted to make acquisitions and/or develop new technologies to add to their existing products and services, the attempt has almost always been unsuccessful or, at best, marginally successful. The culture mitigates too strongly against the addition of other products.

Questions: *Strategic Fit*

- Does the opportunity fit our *corporate strategy* and/or projected strategy?
- How well does this opportunity fit our current *systems, procedures,* and *methods?*
- Is this opportunity in line with the *nature* and *direction* of our business?

Questions: *Implementation*

- Will the organization *reject* this idea?
- Do we understand well the *technology* and *implementation* issues?
- What *track record* do others have in this area?
- Are we *dependent on outsiders* for parts, expertise, services, etc.?
- What are the chances of a raw material shortage?
- How many *new processes and approaches* does this require of us?

This is a checklist of areas that contribute to the strategic fit and implementability of an opportunity. An evaluation of each will complete the framework necessary to assess opportunities.

Checklist

Strategic Fit	*Implementation*
Technology	Current production
Product/service	Current people
Customer base	Degree of control/influence
Market segments served	Current culture
Production needs	Acceptance of change
Distribution	Need for a champion
Method of sales	Financing needs
Resources	Image
Size/growth	Access to outside needs
Profit/return	Legal/regulatory
Direction	Precedent

Making Subjective Judgments Objectively

Many of our clients have noted that costs and benefits seem to be quantitative measures. It's easy to draw this conclusion, but it's not always an accurate one. For example, one can produce quantifiable cost figures in terms of people, equipment, marketing, and so on. In terms of benefit, however, return/profit and market share may produce tangible measures, but the other factors are meant to point up much more qualitative considerations. We have found that too few organizations pay attention to issues such as safety and morale when considering opportunities. An approach that looks great for the organization but not so good for some people within it can turn out to be a disaster rather than an opportunity. Moreover, we find the fallout category of particular importance. By "fallout," we mean the likelihood of the opportunity having repercussions—positive repercussions, one would hope—on other aspects of the business.

Similarly, strategic fit can produce some quantitative measures of resources required for size/growth goals, but it also provides qualitative assessments of, for instance, how well the opportunity fits the current customer base, how well the organization's technology can be adapted, or how well the opportunity fits the direction of the business. The implementation category is meant to be a highly qualitative one in that you should be looking at how well the opportunity meets the current culture, your company's desired image, the need for internal champions, and so on.

While many people are apt to say at the outset, "This seems like a highly subjective assessment system," they soon learn the true point: *there is no substitute for management judgment.* It is not our aim to provide such a substitute, but only to provide an orderly method for arranging facts—both quantitative and qualitative—as well as opinions, so that management can make intelligent decisions about them. We know of no royal road to achieve this, but we do know that when a process exists that makes facts visible and does not give extra weight to the source of the facts, the origin of the opinions, or the decibel level at which they're presented, then that process can be instrumental in helping management to make objective choices. We are making subjective judgments every day as managers, and we *should* be. That's what we're being paid for. If subjective judgment weren't necessary, then there'd be no need for humans on the job at all.

No Substitutes for Hard Work

The more specific the criteria, the more effective the assessment step will be. That's why we encourage clients—and readers—to take the considerations raised by the questions and make them as specific as possible. For example, when looking at a cost that involves legal expense, the most accurate estimate of legal costs should be delineated. If there is a prestige benefit to be gained from the opportunity, that benefit should be spelled out as clearly as possible. The strategic fit relevant to current market segments served should be explored. And the relative acceptance of change in order to implement the opportunity needs to be specified. What usually happens in this activity is that differences of opinion arise, and that's what the process is designed to underscore on paper at this juncture, before money, time, and reputations are being spent on implementing the opportunity.

What one hears at this point is, "What *evidence* do you have for saying that?" "What *precedent* can you cite that makes you feel that way?" "What is the *basis* for the revenue projection you're making?" "Why do you think we'd have difficulty in getting people to *buy into* this approach?"

In other words, although qualitative and often subjective elements are under discussion, the emphasis is on evidence, observable behavior, validity of projections, and reasons for various assumptions. These kinds of questions can be answered by ordering both objective and subjective information into a system that allows it to be evaluated in a high-quality manner. Obviously, responses like, "Because it's always been that way," and, "Because I just *think* so," and, "I don't have to explain it to *you*," are not acceptable in this framework. Consequently, the assessment step becomes the great equalizer; it has proved to be effective in deciding which opportunities to develop as well as when and how to make recommendations to those superiors and/or peers who will be instrumental in the opportunity's acceptance.

We've stated repeatedly up to this point that the process is dependent on as much hard work as the user wishes to invest. In other words, the more *mental* the work—the more disciplined the approach to each step—the more accurate and useful the process will be. This is nowhere more apparent than in the assessment step. For example, if one merely wished to make random assessments—such as, "Well, I think the difficulty of implementation will be fairly high," or, "I don't think the cost here will be too bad and the return should be pretty good"—then the results will reflect exactly that kind of inaccuracy and lack of work. On the other hand, if you take the time to go through the checklists, modify them to your personal needs, and make careful determinations about the degree to which the opportunity at hand meets or does not meet the considerations on your checklists, then the outcome should be accurate and useful.

TRACKING OPPORTUNITY ON THE ASSESSMENT GRID

Managers need some system that will provide a road map for discovering the information needed to assess opportunity. We've found that there's not so much a lack of volition among managers, as a lack of direction. Therefore, the reason for the various categories, their subsets, and their modifications, as well as for the grids on which all this information will soon be charted, is not to maintain neatness or documentation (although these by-products are benefits in and of themselves). Instead, these approaches are specifically aimed at providing *direction,* so that the hard mental work will be less onerous. You might say that this is our attempt to modify your perception. Think of the difficulty of trying to assemble a child's bicycle or a new stereo component system with the directions missing. The physical work remains the same as it would be with the directions in front of you, but the mental work is much more taxing when the assembly process is a mystery.

It's not generally the ambiguity about *content* that gives managers problems, but rather any ambiguity surrounding *process* and *format.* Content is ambiguous by nature—there will always be uncertainties that managers are forced to confront as part of their jobs—but there is no reason for a management *process* to also contain ambiguities. The entire innovation process, and particularly this assessment step,

Figure 4–1. Opportunity Assessment: Cost and Benefit.

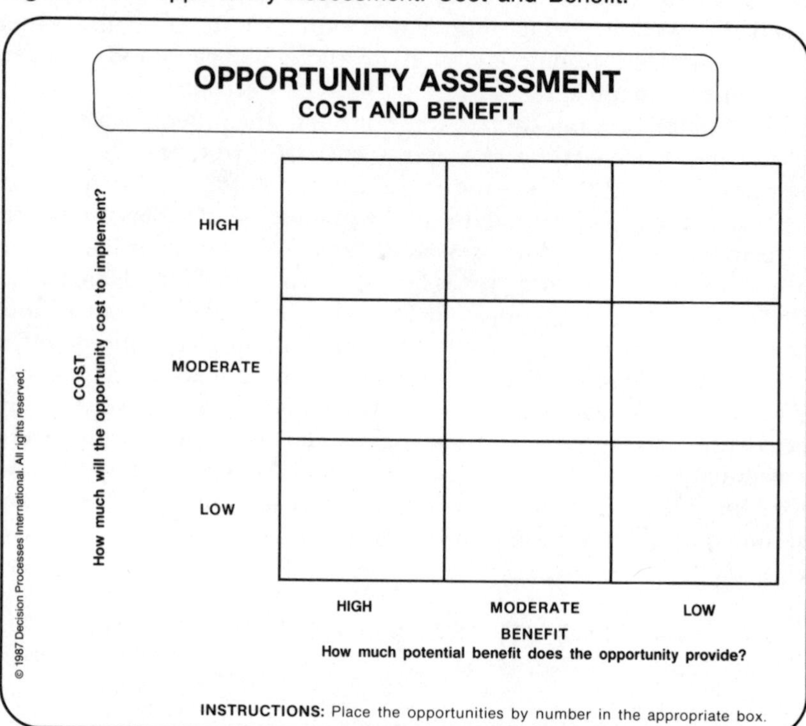

is designed to minimize and/or eliminate ambiguity in the pursuit of innovation. If the process is clear, we find that managers can be much more effective in dealing with any ambiguities surrounding the matters at hand.

The goals of the assessment step are to evaluate opportunities developed in the search step, to evaluate opportunities that "strike from the blue," and to evaluate, when and where appropriate, existing methods of doing things. As a result of the assessment step, opportunities that produce the most benefit within acceptable risk limits and seem to fit the organization's direction and implementation ability can be developed still further. The assessment step can be seen as a way of filtering, of very carefully setting priorities, so that opportunities that should receive very specific attention can be determined. To make them as visible as possible, to shine a spotlight on your star opportunities, we've developed an assessment grid to assist you in this pursuit.

In Figure 4–1, cost is reflected on the vertical axis as high, moderate, or low, as is benefit along the bottom axis. These are obviously very general classifications, and we encourage you to make them more specific, if possible.

For example, the cost classifications can easily be broken down into specific amounts of money. Many of our clients, depending upon the issues at stake, define low cost as under $100,000, moderate cost as $100,000 to $500,000, and high cost as above $500,000. Obviously, in this example we would be talking about opportunities that represent major changes, and therefore, major investments. Opportunities related to a smaller issue—departmental, personal, small-business—might be charted with low representing $5,000 and under, moderate $5,000 to $20,000, and high over $20,000. It's very important to establish what these parameters are, since various people will probably be involved in the actual assessment. You want to avoid one person thinking that low cost represents anything under $100,000 and another person thinking that low cost represents anything under $5,000. While such differences of opinion often reflect different perspectives in the organization, it's relatively easy to achieve consensus as to what low, moderate, and high should actually represent when particular issues are being discussed. In fact, one of the greatest reasons that some people's pursuit of an opportunity seems dangerously reckless to others is a distinct difference in perception about what constitutes high and low cost, and consequently, what constitutes high and low risk. The cost dimensions can often be established from an investigation of operating budgets, levels of authority, expectations of management, and precedent. No matter how they're established, cost parameters must be agreed upon beforehand so that the assessment step employs criteria agreed to by all parties concerned.

Similarly, high, moderate, and low along the benefit axis can be further specified using the checklists discussed earlier in this chapter. Again, neatness *does not* count, objectivity *does.* Therefore, when the actual opportunities are charted on the grid, you may wish to make notations about why you've assessed them where you have. For example, if an opportunity reflects high benefit primarily because of the fallout factor, you may wish to footnote that entry with the comment that the benefit should increase sales of related products as well as contribute to the image of the company as being on the leading edge of its technology.

For simplicity's sake, and to keep the grid within manageable bounds, we recommend that opportunities be plotted by number. Therefore, the opportunities generated in the search step should be placed on a "master list." This is simply a list of the opportunities that have been generated, annotated with specific descriptions of each. *The assessment step works best when the opportunities being considered are as specific as possible.* For example, try to assess the cost—high, moderate, or low—of "expanding customer service hours to two additional nights a week in three of our stores in Chicago." Benefit is also much easier to track, the more specific you're able to be.

The master list of opportunities should be numbered sequentially so that those numbers can be plotted on the assessment grid. By keeping the master list of opportunities juxtaposed with the grid, you can easily track where you are at any given time.

THE ASSESSMENT GRID AT WORK

Let's suppose that you're the general manager of an auto dealership. The opportunity that you've come up with—for the moment, it doesn't matter what its source is—is to expand service hours so that customers can now have their cars serviced from 9:00 a.m. to 3:00 p.m. on Saturday, in addition to your normal Monday–Friday 8:00 a.m.–5:00 p.m. service. This item is numbered "one" on your list.

Your criterion for low cost is anything under $10,000; for moderate, from $10,000 to $25,000; and for high, anything over $25,000. You estimate that you would have virtually no additional costs in the areas of insurance, equipment, materials (paid for by the customer), marketing, or upkeep. Additional utility costs would be minimal, and the sales department is open on Saturday anyway. The sole cost of any magnitude will be the overtime salaries for the mechanics and one assistant service manager. You estimate that this cost will amount to about $48,000 a year. Consequently, you know that for cost purposes, you are in the upper left-hand portion of the grid.

Assessing benefit, you feel that Saturday hours would bring you some significant gains. Your current workload is always far in arrears, and your current loyal customers are concerned about it. Moreover, the new-car sales area is concerned about what it can legitimately promise new customers about prompt service. Finally, you see a tremendous increase in your revenues because you'll be able to attract people who didn't buy their cars from your dealership but who could be drawn in by the Saturday service hours. You're thinking especially of two-income families who don't have the luxury of dropping the car off during workdays.

Repairs have long been one of the most profitable parts of your business, since all materials are paid for by the customer on a cost-plus basis and labor has a substantial markup. You estimate that you could handle fifteen customers on each Saturday. Since the average service bill comes to about $150, you see the ability to gross over

Figure 4–2. Example of Opportunity Assessment: Cost and Benefit.

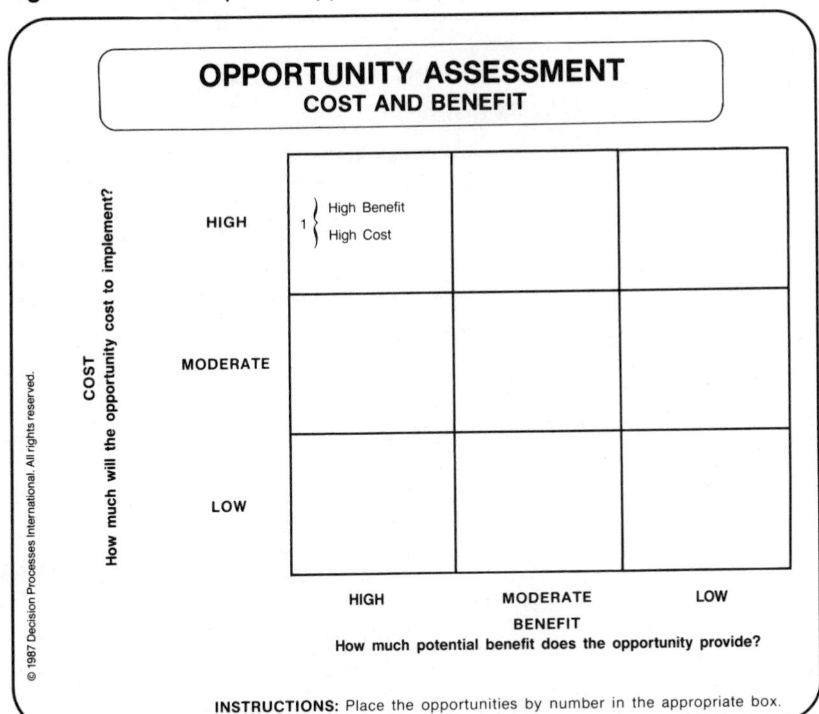

$100,000 annually with new Saturday hours. You also believe that the ability to reduce complaints from customers who otherwise couldn't bring their cars in for prompt service will add considerably to the dealership's prestige and quality reputation. The morale of the mechanics should be helped—they've been asking for more overtime—and the new-car salespeople should be more confident about what they can promise. Finally, you think it would be fairly easy to expand if things go well, perhaps even hiring additional mechanics and lengthening Saturday hours to 8:00 a.m.–5:00 p.m., as they are during the week. In this manner, you might be able to handle as many as twenty-five cars a day six days a week, with commensurate profits.

You believe that this combination of financial and nonfinancial factors rates a "high" on the benefit scale. With high benefit and high cost, you place the number "one," representing this opportunity, in the upper left-hand box in the grid. (See Figure 4–2.)

Now let's examine the strategy and implementation part of the grid. On the vertical axis is implementation ability. The question to ask here is, "How difficult will it be to implement this opportunity?"

Figure 4–3. Opportunity Assessment: Strategy and Implementation.

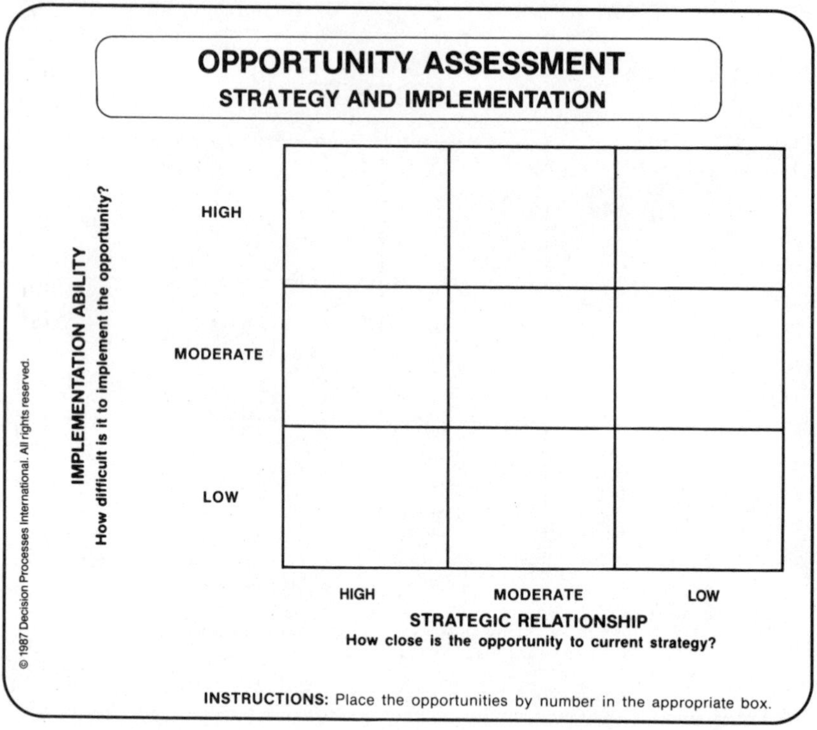

Again, you'll need some agreement from your colleagues as to what constitutes high, moderate, and low, using the factors cited earlier in this chapter, in order to make an objective assessment. For example, low implementation difficulty might mean that the idea can be absorbed into the current operation using existing resources and procedures with hardly any disruption at all. On the other hand, a high rating could mean that various influential people will have to be convinced, some standard operating procedures might have to be redrawn, resources might need to be shifted, and company culture might require a good kick in the pants.

Along the bottom axis we have strategic relationship. Asking, "How close is this opportunity to our current and projected strategy?" a high strategic fit might mean that the idea fits cleanly and neatly into the existing statement of strategy and beliefs. A low strategic fit would mean that the idea is at best disassociated from, if not antithetical to, existing strategy. As we mentioned earlier, some "simple" ideas, such as reorganizing a department or changing a method of sale, might automatically be a high strategic fit because they concern largely tactical issues that have no strategic bearing. On the other hand, some larger issues—such as moving into a new marketplace, introducing a new product, changing pricing—would almost certainly require careful analysis for strategic fit.

Let's go back to our example of the auto dealer. The general manager feels that the new hours can be implemented with current people working overtime, which is something they've requested anyway. The change should therefore be well accepted by the people who have to implement it and should be received with open arms by the customers. There is no legal problem involved, nor is there a cultural difficulty. The financing to begin the new hours can be more than covered by the existing budget. Quite rapidly, the endeavor should become self-sustaining. For these reasons, the general manager would rate difficulty of implementation of the new Saturday hours as "low," in the bottom left-hand corner of the grid.

As far as strategic fit is concerned, the general manager knows that the president of the dealership is anxious to have the organization viewed as a full-service, customer-oriented business. Moreover, the car manufacturer itself is currently stressing longer term six-year/60,000-mile warranties for its products. The expanded hours would seem to fit in well with the expanded service being promised by the automaker.

The products and services offered are the same as those being offered during the week, and the markets served are also identical. There is no need to change anything about production, the way parts will be sold, or the resources needed to provide the service. For these reasons, the general manager evaluates the strategic fit as "high," also in the bottom left-hand corner of our grid.

Notice that in our example with the auto dealership, the questions we've been asking in terms of cost, benefit, strategic fit, and implementation come from the checklists provided earlier in this chapter. We've used a somewhat basic example so that we can all relate to it; nonetheless, we were able to become highly specific simply by referring to the checklists and the general questions also mentioned above. As you further modify and refine these process questions and checklists for your individual environments, you'll find it even easier to determine where opportunities fit in your assessment grid.

Our auto dealership general manager has now made an assessment using the four basic factors and has arrived at the following: In terms of cost, the general manager has rated the opportunity to expand service hours as "high," and "high" also in terms of strategic relationship. Note that the number for this opportunity, "one," has been listed on the grid *twice*. The first time, it was listed in the upper left, reflecting high cost and high benefit. The second time, the number

Figure 4–4. Example of Opportunity Assessment: Strategy and Implementation.

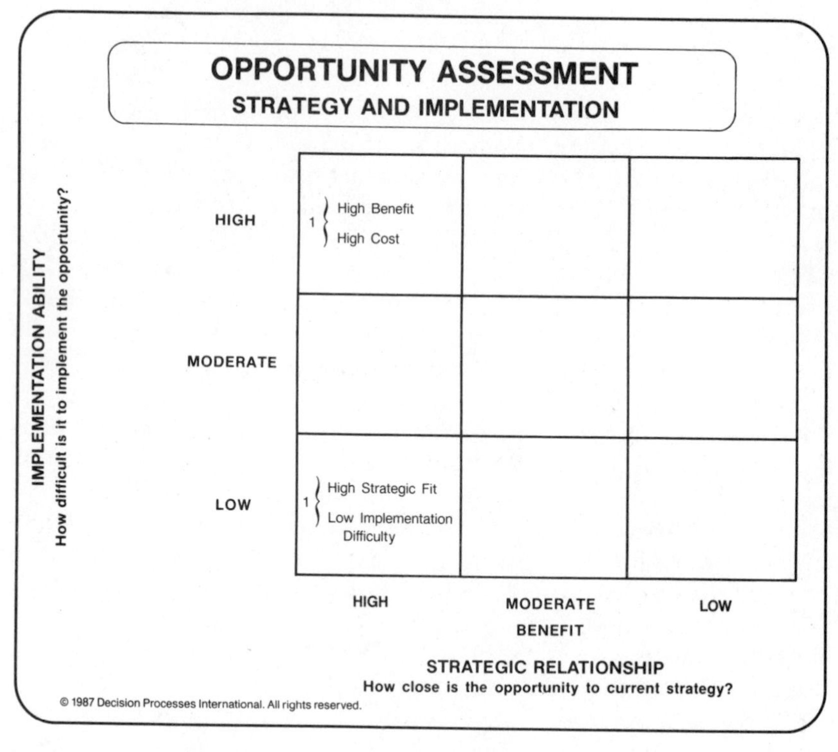

"one" was listed in the lower left to signify low difficulty of implementation and high strategic fit. While sometimes a given opportunity can be plotted in the same place after both analyses, it's more common to find it appearing in two different places. The assessment grid is not designed to produce some magic, "clean" answer. Rather, it is designed to reflect exactly what the opportunity provides in terms of the factors being assessed. How, then, are you to make a judgment about whether or not to pursue a given opportunity when it appers in two different places on the grid, as does our example? The answer is in the final phase of the grid, which provides you with an opportunity assessment ranking.

This ranking is not intended to supersede management judgment. Its purpose is to provide *clarity,* to remove the ambiguity that might get in the way of the *process.* Therefore, the ranking system merely points out what tends to be best, that is, which opportunities tend to look better than others at this still fairly early stage. It remains up to the manager, however, to determine which opportunities should

Figure 4–5. Opportunity Assessment Ranking.

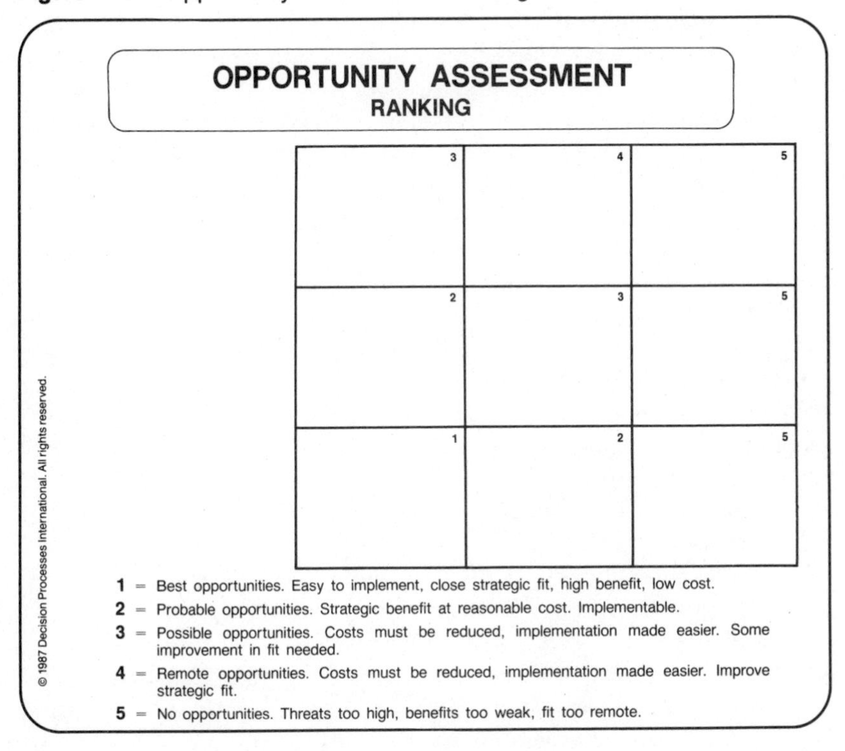

go on to the development step. As many of you have no doubt anticipated, this is because managers can find ways to reduce high cost, enhance low benefit, ameliorate difficulty of implementation, and improve poor strategic fit. None of these things can be readily accomplished, however, unless we can see clearly how worthwhile such improvement efforts can be, given the amount of payoff we can reasonably expect.

By examining the opportunity assessment ranking system, we can see that there are five possibilities for opportunities appearing in one or two of the nine boxes of the assessment grid. We have keyed these five possibilities by number.

1. This one box of the grid represents what are obviously "best" opportunities. It means that cost is low while benefit is high, implementation is fairly easy, and strategic fit is high. This is the best of all possible worlds, and any opportunities finishing in this box after both sets of assessments should definitely by carried over to the next step of the process, opportunity development.

2. These two boxes represent "probable" opportunities. They would provide reasonable benefit, with reasonable cost. They fit well enough into the organization's direction and can be implemented without too much difficulty. These are opportunities that will require fine-tuning, but should almost always be pursued and taken to the next step.

3. These two boxes represent "possible" opportunities. Cost might have to be reduced, and implementation made easier. Some improvement in strategic fit will be needed. Possible opportunities need more than fine-tuning; they require some careful work. Nonetheless, they may still be excellent candidates for developing further.

4. This box represents a "remote" opportunity. By "remote," we mean you'd better have a very good reason for deciding to develop it further. Costs are high, while benefits are moderate. Although strategic fit is moderate, ability to implement will be strained. Remote opportunities are usually long shots, which, of course, can occasionally pay off. But if you decide to develop them, you need to do it with your eyes open to the attendant risks. The next step, development, will be crucial.

5. These three boxes, the entire right side of the grid, are red flags. These are not opportunities, no matter how good they might have looked emerging from the search step. Threats are too high, benefits are too weak, the fit is too forced. We find that the chances of shipwreck are all too high for those who attempt to sail in these waters. The ideas that ended up in these boxes are worse than a remote possibility, which could at least be managed to a more favorable outcome. These are strictly gambles, rolls of the dice; they are highly dangerous.

Now let's go back to our auto dealership general manager. As you can see in Figure 4–5, his opportunity would appear as a "best" in one box and as a "possible" in another. It's a "possible" because of one factor: the relatively high cost that has been attributed to its implementation. There are two ways to view this situation. With three of the four factors ranked "one" and only one factor ranked "three," it can be seen as a legitimate and fairly safe opportunity to pursue to the development step. And that happens to be quite true. A second way of looking at it, however, is to notice that the high cost was arrived at without any consideration of the return. Using the figures previously generated, the general manager can see that the operation will make a profit—a fairly handsome gross profit of about 50 percent—soon after start-up. Consequently, the general manager feels completely justified in changing the original high cost assessment to "low." Now, the opportunity to expand service hours to Saturdays is moving toward the lower left-hand corner.

INTERPRETING THE GRID AND FURTHER RECOMMENDATIONS

We've made this example deliberately neat. There will be times when it's not so easy to make a judgment, and when an opportunity can't be easily placed in the desirable number-"one" box. What happens if, on the first pass, through cost and benefit, an opportunity winds up smack in the center of the grid—ranked "three"—and on the second pass, through strategic fit and implementation, it finishes in the middle box in the left-hand column—ranked "two"? Well, the point is not to go through the assessment step with the intention of getting each opportunity squarely and perfectly affixed in a single box. In this last example, you would have an opportunity that finished in both a probable and a possible category. It is now up to your *judgment* as to whether that opportunity should be further developed. Don't forget, the development step is also a "paper" step, one that can be completed without a great expenditure of time, money, or resources. And certainly, the development step can be completed without an investment of personal or corporate reputation. If time and focus are limited, then you will tend to develop only those opportunities that cluster toward the bottom left of the grid. When time is not a critical factor, however, or when there needs to be an exhaustive examination of opportunities—as we have found to be the case in highly competitive organizations where innovation can mean the difference between profit and Chapter 11—then you can take a more studied look at potential opportunities that appear elsewhere on the grid.

Up to this point, the assessment step has helped you deal with the relative worth of opportunities. It has allowed you to compare them against each other, and perhaps against existing deployment of assets and resources, thereby enabling you to make some reasoned conclusions about their attractiveness. In the next step of the process, the development step, you will be able to take each particular opportunity that has appeared to be attractive after the assessment step and project it out, uniquely and singly. Consequently, there is no need to do that at this point; in fact, doing so would simply get in the way of the assessment process.

We do, however, have some simple recommendations that we feel are applicable in most circumstances.

Recommendations

1. If an opportunity finishes both assessments ranked "five" for each one, eliminate it. Pursuing these opportunities is worse than trying to draw to an inside straight. It's probably more akin to skating across extremely thin ice with a 500-pound packpack and a flamethrower pointed at your feet. It is a blueprint for disaster. Disaster often enough pursues you, there's no need for you to pursue it.

2. Eliminate opporunities that are ranked "four" in both assessments, *unless the opportunity can provide the organization with an absolutely unique competitive advantage if successfully implemented.* Since you will still use the fail-safe development and pursuit steps to further examine outcomes, you could pursue the opportunity on those grounds. A "four" ranking, however, *is* the inside-straight phenomenon. Even most poker experts recommend breaking up the hand rather than trying to draw to an inside straight.

3. For the reasons cited above, do not pursue any opportunities that finish one assessment ranked "five" and the other ranked "four."

4. Develop those opportunities that appear both times ranked "three" only if you feel confident that you can mitigate the circumstances that have resulted in it being assessed as only "possible." A "three" ranking is an excellent chance to use your visceral and intuitive feelings. *The more experienced the manager, the safer the manager is in choosing to develop "possible" opportunities.* The more inexperienced or new the manager, the less safe that manager is in attempting to go with these opportunities. Again, the fail-safe steps of development and pursuit lie ahead, but we're also concerned with the investment of time and energy. Usually, but not always, those opportunities finishing *in the middle* of the grid are somewhat safer to pursue than those that finish both times in the upper left-hand corner of the grid.

5. For the reasons stated above, eliminate opportunities that finish one assessment ranked "four" *or* "five" and the other ranked "three."

6. In general, you should develop opportunities that finish both assessments ranked "two," or one ranked "two" and the other ranked "three." The latter kind of opportunity is usually the one that deserves the most *critical* scrutiny in the development step. It is usually a good idea, however, to develop them, particularly those that finish in the left-hand column ranked "two" and "three," since these boxes represent high benefit.

7. Any opportunities that finish both assessments ranked "one" should obviously be pursued with a vengeance. The development step, in this case, should be used to explore the power of their benefits. Similarly, any opportunity finishing one assessment ranked "one" and the second assessment ranked "two" or "three" should be pursued. These should be pursued with an eye toward ameliorating the difficulty and maximizing the opportunity. (To reiterate our warning, we recommend against pursuing any opportunity that finishes one assessment ranked "one" and the second assessment ranked "four" or "five," *unless you can provide a tangible and demonstrably powerful reason for so doing, such as unique competitive advantage.*)

8. What of the opportunity that your "gut feeling" won't let you abandon, despite its finishing the assessment step ranked "four" and/or "five"? You must demonstrate that you can reasonably "drive" it toward the lower left-hand corner of the chart by enhancing the benefits, lowering the cost, improving the fit, and easing the implementation. If you *can't* do that, you're letting emotion—or selfishness—rule over logic.

We now have a "short list" of attractive opportunities. The development step, and ultimately the pursuit step, will provide further refining as well as a fail-safe mechanism to protect you. This does not mean, however, that the assessment step should be looked at lightly. On the contrary, we find that it's a unique way to objectively determine which opportunities can produce the most for the organization and/or the individual. It tends to eliminate volume, emotion, politics, and competitive fear as the bases for deciding which courses of action to pursue. It helps you to carefully order a diverse variety of information, assumptions, and opinions at your disposal.

In keeping with the process, this diversity of fact and opinion should not be looked at as threat, but as opportunity. The assessment step is the vehicle we have found most advantageous for ensuring that this opportunity is exploited.

Chapter 5
OPPORTUNITY DEVELOPMENT: THROUGH THE MINE FIELD

We are now ready to take the highest potential opportunities from the assessment step and develop each of them, prior to implementation. The development step is the preparation for implementation, or what we will call the pursuit step. There are several results that may emerge from the development step:

1. An opportunity will prove to be highly beneficial and well within acceptable risk. You will take steps to maximize its potential benefit, and it will enter the implementation process in the pursuit step.
2. Previously unrecognized risks in a given opportunity will become visible. You will evaluate the risks in terms of the opportunity's potential benefit and decide whether to proceed to the pursuit step or not.
3. Available information is so sketchy that it will be impossible to complete the development step with any accuracy. You must obtain further information prior to proceeding to the pursuit step.

As you can see, the development step serves both as a preparation phase for implementation *and* as an additional filter that enables you to continue to cull the opportunities that don't provide sufficient benefit. The middle portion of the innovation process—the assessment and development steps—can be looked upon as "destructive test" mechanisms. That is, rather than too easily supporting opportunities derived in the search step, these ensuing steps provide critical analyses of those opportunities. They enable you to make certain that all required information is present, and that the information has been evaluated with scrutiny and diligence. For an opportunity to qualify for the final pursuit step, it must have satisfied some very rigorous testing along the way.

Our belief is that this testing is crucial, because all too often people tend to jump aboard a bandwagon to support an idea that looks good on the surface. By providing a deliberately critical evaluation stage, these two steps ensure that even the most ardent supporters have verified assumptions and validated the evidence that supports their conclusions. Seeking and exploiting opportunities for the future may be far from a sure thing, but it is possible to consciously minimize risk by being your own devil's advocate.

In assessment, opportunities were examined relative to each other and relative to set criteria. In development, those opportunities that emerged from assessment are evaluated singly. Depending on the issues at stake, this is a process that can involve one individual spending less than an hour on a particular option, or a committee taking several days to detail the pros and cons of a given opportunity.

MANAGING RISK

Why is this further evaluation necessary after making the evaluations involved in the assessment step? For one reason, development is a fail-safe step. Whereas the opportunities in the assessment step were bathed in a floodlight, in development the details of each particular opportunity are explored with a spotlight. For another, we found it important to include some redundancy in evaluating opportunities. Because of what is riding on any given opportunity, and because of the attendant risk in any kind of innovative effort, we included this extra evaluation step as a guarantee to provide the safest possible approach to innovation. In Chapter One, we talked about the need to take *prudent* risk with innovating. We found that this redundancy assures that such prudence *is* undertaken.

A third and final reason for the development step is that it raises the critical factors that will be the key to implementation in the pursuit step. All too often, plans and implementation based on generalities are undertaken, without a careful understanding of specific details. These critical factors raised in the development step become

Figure 5–1. Cause and Effect: Problems.

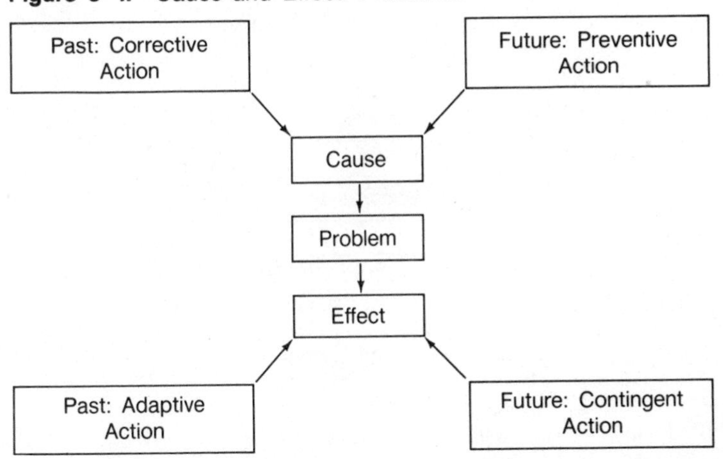

the basis for discrete and specific actions that are included in the action plan of the pursuit step.

ACTIONS TO TAKE

The final step in the innovation process—the pursuit step—will concentrate on special actions to remove obstacles to the effective implementation of the opportunity. This activity includes solving potential problems. The fundamental premise in effective problem solving is a simple one: *you cannot remove a problem without removing its cause.* You can *adapt* to a problem's effects without knowing the cause, but you can never completely remove the problem unless you know the cause and take action against it. (Similarly, you cannot *promote* an outcome without knowing *its* cause.)

Note that there are several variables to consider when deciding whether to take *adaptive* or *corrective* action. These variables include cost of the various actions, inconvenience caused by the actions, time available to take the actions, and ultimate severity of the effects on the innovator. When trying to avoid or mitigate future problems, we take preventive actions against causes and contingent actions against effects. "No smoking" signs are meant to prevent the cause of a fire (careless smoker) and sprinklers are intended to reduce the effects if the problem occurs.

All of these actions, however, deal with problem solving. Since we are talking about innovation and opportunity, we need to add two

Figure 5–2. Cause and Effect: Opportunities.

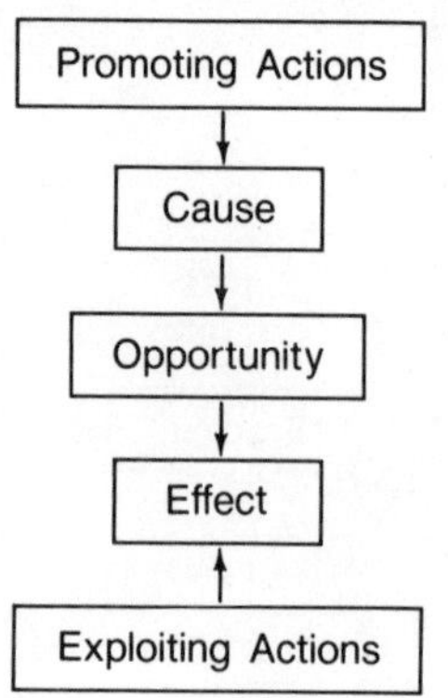

types of action to our list. Those actions are *promoting* and *exploiting*. (See Figure 5–2.)

Promoting actions also address causes and, like preventive actions, address those causes in the *future*. But if preventive actions seek to keep the cause of a problem from reoccurring, promoting actions seek to ensure that the cause of an opportunity *will* occur. There is nothing about innovation that should be left to chance. Consequently, as critical factors emerge that could spell the difference between the success or failure of an opportunity, it is important to promote those critical factors that will cause the best possible outcome. Similarly, exploiting actions seek to maximize the benefit that is achieved from the effects of the opportunity. You might call this a "capitalizing" action. As contingent action seeks to mitigate the effects of some potential problem, exploiting action seeks to further stimulate the benefit from any of a potential opportunity's effects that may occur.

For example, with the deregulation of the telephone industry, many alternative publishing companies have started to produce their own yellow pages. At least one of the critical factors in the success of these alternative yellow pages is the inclusion of as many businesses and services as possible, so that the new book is seen as a viable alternative to the standard telephone company yellow pages. To promote the likelihood of this occurring, many of the alternative yellow pages publishers provided free listing for advertisers in the initial edition of their books. This was seen as a way to get as many businesses involved as possible and to demonstrate the credibility and utility of their books from the outset. This decision to provide free advertising in order to launch the books was a key *promoting* action. Still another promoting action was to provide the books for free in public places. Many publishers sought and received permission to provide displays with free copies in such places as local post offices. This greatly enhanced the distribution of the books and was another fundamental promoting action.

These actions were taken to stimulate use and to get the books into the hands of potential users. Now, what happens if the desired effect develops: people actively begin using the alternative yellow pages directory? Well, there should be several desirable and opportunistic effects. These effects will include applications for inclusion in the next printing of the book; the ability to charge competitive rates for future advertising; the credibility and financial support to expand circulation of the book into other communities; and the demonstrated usage that would justify outside financial support for expanded marketing activities. It's not sufficient, however, simply to sit back and wait for this to happen and to congratulate yourself if, indeed, it does. There should be plans in place—actions considered—that will exploit these positive effects of the opportunity. This is where exploiting actions enter the picture.

In our example, exploiting actions could include any or all of the following:

- Offering advertising contracts to future advertisers for periods of two, three, or more years at discounted rates.
- Establishing a modest charge for the book itself—a fee that would be tolerable to almost any user who is happy with the book, yet would help to defray some printing expenses
- Establishing a subscription service so that users could order yellow pages for those communities most useful to them at a reduced rate, thereby alerting potential advertisers to the demographic appeal of the book

In the opportunity development step, we will therefore be carefully considering cause and effect and their relationships to our opportunity. That's why the development step raises critical factors specific to each opportunity that weren't necessary in the prior assessment step. The development step will examine what we call the best-case and worst-case scenarios. That is, we ask that you take each opportunity individually and project it into the future, describing the best that you can expect to occur, and the worst. While "worst" may seem like an odd choice of word for an opportunity, we find that this is not at all the case. Opportunities have a range of potential effects on the individual and/or the organization. Sometimes it's assumed that those effects can only be highly positive and can occur only in a fairly narrow range. But, in fact, these effects can occur across a fairly broad range, sometimes so broad as to include a decline from the current status quo.

By looking at best-case and worst-case scenarios, we found that managers are able to proactively manage the innovation process, using promoting and exploiting actions to try to ensure the best-case scenario and preventive actions to try to avoid the worst-case scenario. Depending upon the appropriateness, realism, expense, and control of those actions, a manager is then able to determine how feasible it is to take the opportunity into the final pursuit step. In other words, what confidence do you have that you will be able to avoid the worst case and produce the best case? This is a key question to ask while you are still working out the opportunity "on paper," but it's too late to ask once monies have been allocated, people have been moved, and reputations have been placed on the line.

To help in the evaluation of best- and worst-case scenarios, we've prepared some checklists for consideration. The best-case/worst-case evaluation is a subjective judgment—a management judgment—but one based, again, on objective criteria. We've chosen seven areas to review in determining what best-case and worst-case results you can reasonably expect. These are especially useful to review with others. In many cases, teams deliberately bring in outsiders—individuals who will have no involvement in the ultimate implementation of the opportunity, and who can be especially objective in helping to review these factors and areas in terms of the best- and worst-case projections. You might seek to add other factors and other questions, depending upon your organization, your own style, and the culture in which you work. We encourage you to do so.

Critical Factors Checklist

Technical

- Availability of requisite scientific/technical skills
- Adequacy of research resources
- Quality and quantity of support personnel
- Probability of technical success and validation
- Government/regulatory position

Timing

- Research completion versus market need
- Market preparation and development
- Known and assumed competitive actions

Stability

- Durability of the market
- Chances for a dominant or preeminent position
- Probability and impact of "down" markets

- Stability of largest projected users
- Volatility of the approach

Position

- Impact on other products and services
- Impact on overall credibility
- Ability to assume a rapid leadership position
- Ability to facilitate other opportunities

Growth

- Short-term market potential
- Long-term market potential
- Impact on market share
- Promotional requirements to launch
- Promotional requirements to sustain
- Adequacy of present distribution systems
- User view of cost versus value
- Applicability to current customers
- Servicing requirements
- Impact on reputation

Production

- Production capabilities
- Logistical demands
- Expertise required

Financial

- Projected return on investment
- Impact on earnings per share
- Degree of new capital outlay required
- Projected total cost
- Contingencies required

PROJECTING RISKS AND REWARDS

Reviewing these factors, you should list (see Figure 5–3) your projections about the worst-case scenario and the best-case scenario in terms of yield and productivity from the opportunity. Since entrepreneurs redeploy assets and resources to improve yield and productivity, the development step will enable you to determine if you're engaged in a truly entrepreneurial activity. For example, a best-case scenario in terms of yield and productivity for a hypothetical company might look something like this:

- Market share gain from current 13 percent to estimated 16 percent

057599

Figure 5–3. Opportunity Development Worksheet: Impact Analysis.

OPPORTUNITY DEVELOPMENT WORKSHEET

State the Opportunity/Concept: ______________________________

WORST-CASE SCENARIO YIELD/PRODUCTIVITY	BEST-CASE SCENARIO YIELD/PRODUCTIVITY
No increase in market share	Market share gain to 16%
Reduce personnel to nine	Reduce personnel to seven
Appeal to 50,000 new users	Appeal to 250,000 new users

RISK/REWARD ANALYSIS

- Compared to the status quo, how much positive impact will the best-case scenario have?
- Compared to the status quo, how much negative impact will the worst-case scenario have?

-5	-4	-3	-2	-1	0	+1	+2	+3	+4	+5
NEGATIVE IMPACT					*STATUS QUO*					*POSITIVE IMPACT*

- What critical factors will determine whether we achieve the best-case scenario?
- What critical factors would bring about the worst-case scenario?

(List these on next worksheet)

© 1987 Decision Processes International. All rights reserved.

- Ability to cover northeast territory with seven salespeople rather than eleven
- Appeal to a user group not currently being reached is estimated at 250,000 people

Examples of a worst-case scenario might be:

- No increase in current market share
- Reduction in salespeople covering northeast territory from eleven to no fewer than nine
- Estimated numbers on access to a user group not currently reached are unknown, but 50,000 would be the minimum

Having listed the projections that add up to both the worst-case and best-case scenarios, you're now in a position to list the critical factors that will tend to bring about the best case and the critical factors that will tend to bring about the worst case. Using our example above, let's say that the critical factor that will enable us to reduce the number of salespeople in the northeast territory is the rapid assimilation of the new computerized ordering system by the top performers. The critical factor underlying our chance of gaining market share will be our ability to smoothly implement our new computerized ordering process before the competition does; in fact, our beating them to the punch, and the degree to which we do so, is the all-important factor in this area. The critical factor in reaching the 250,000 people who constitute a new user group is the acquisition of accurate mailing lists, which will help us determine where the users are and how to reach them.

The worst-case critical factors will sometimes be the converse of the best-case critical factors, but sometimes they will be quite different. For example, a critical factor in our inability to decrease the number of salespeople from eleven to any less than nine might be the sales force's inability to rapidly assimilate the new computerized ordering system. Another critical factor in this area, however, might be the perception by the sales force that we are looking to replace all of them eventually; for this reason, they may be deliberately reluctant to use the new system to its maximum potential. A critical factor in our not being able to increase the market share might be our delay in introducing the new computer ordering system, thereby losing ground to the competition. But it might also be a competitive action that offsets our early advantage.

In any case, the sequence is clear: having reviewed the checklist of factors for consideration, you then detail the best-case and worst-case scenarios that you feel are reasonable to expect from the implementation of this opportunity. The next step is to determine what critical factors will tend to bring about the best-case and worst-case scenarios so that the appropriate actions can be undertaken in the pursuit step of the process. There is, however, one more element in the development step: a risk/reward analysis of what this opportunity can *actually* provide for you and your organization.

RISK/REWARD ANALYSIS

The risk/reward analysis uses a scale that evaluates an opportunity as to its positive or negative impact relative to the status quo. We've arbitrarily used a five-point scale in each direction, with zero rep-

Figure 5–4. Opportunity Development Worksheet: Critical Factors.

OPPORTUNITY DEVELOPMENT WORKSHEET

OPPORTUNITY/CONCEPT: ______________________

List the critical factors that will tend to bring "best-case" success:

Assimilation of new ordering system by top performers

Implementation of new system before competition

Acquisition of accurate mailing lists

List the critical factors that will lead to "worst-case" performance:

Inability of top performers to assimilate new system

Sales force perception that everyone is to be replaced

Delay in introducing new system

Competitive actions that pre-empt us

DOES THIS OPPORTUNITY PRESENT SUFFICIENT BENEFIT WITHIN ACCEPTABLE RISK LIMITS TO TAKE TO THE PURSUIT STAGE? ____YES ____NO

© 1987 Decision Processes International. All rights reserved.

resenting the status quo. The calibrations on the scale are unimportant. You can choose larger numbers, more numbers, or finer calibrations. The point is, however, to evaluate how large a gap exists between the best case and the worst case, between risk and reward.

Let's suppose that following your analysis of best-case and worst-case scenarios, you determine that the best case merits a +4 on our scale. Your evaluation of the worst-case scenario yields a +1. In other words, even in the worst case you will have an improvement over the status quo, and in the best case you'll have a considerable improvement. In this instance, the opportunity should definitely be taken forward

Figure 5–5. Risk/Reward Analysis: Examples of Outcomes.

Situation 1

WC (+1) to BC (+4)

−5 −4 −3 −2 −1 0 +1 +2 +3 +4 +5

Excellent: No risk; anything in the entire range of outcomes will be an improvement.

Situation 2

WC (−3) to BC (−1)

−5 −4 −3 −2 −1 0 +1 +2 +3 +4 +5

Whoops! All risk; any result in this range makes things worse.

Situation 3

WC (−1) to BC (+4)

−5 −4 −3 −2 −1 0 +1 +2 +3 +4 +5

High potential: A minimum of risk management is required.

Situation 4

WC (−3) to BC (+1)

−5 −4 −3 −2 −1 0 +1 +2 +3 +4 +5

Low potential: Nothing here is worth the risk.

Situation 5

WC (−3) to BC (+3)

−5 −4 −3 −2 −1 0 +1 +2 +3 +4 +5

Character test: Do you flip a coin? This is where critical factors play the crucial role.

to the pursuit step, since it's one you literally cannot afford not to pursue. The impact is only positive, and ranges from moderately to significantly positive.

Let's now assume, however, that your best-case opportunity is a −1 and your worst-case opportunity is a −4. In this case, the opportunity should be dropped at this step and not taken forward to pursuit. Yes, it is possible for an opportunity that looked promising in the assessment step to finish up with a negative impact analysis in both best-case and worst-case scenarios. Why is this? Well, sometimes such opportunities were evaluated in the assessment step as "remote," but you chose to take them forward anyway to see if they could be improved. Still other times, they are opportunities that finished in the "possible," or even the "probable," category in the assessment step, but additional information generated in examining them has lessened their attractiveness. This is the reason that the development step is placed where it is—to ensure that opportunities are destructively tested and carefully scrutinized.

Life being what it is, however, it is common to find an opportunity whose best case rates a +3 and whose worst case rates a −2. *It is in these instances that identifications of critical factors, and eventual actions required by those factors, are especially crucial.* Where there is clear benefit *and* clear risk, the promoting and preventing actions will be decisive in enhancing one and minimizing the other. With this type of opportunity, the development step is the juncture at which you make the decisions as to whether the risk involved is worth the potential benefit, whether other actions need to be examined to try to reduce the risk, and how dependent—vulnerable—you are to factors beyond your control.

Risk/Reward Analysis Definitions

- −5 equals disastrous results; business, jobs, morale, and/or image will all suffer severely; could represent failure of a small business, "black eye" for a large business; severe risk.
- −4 equals substantial risk; major disruptions to organization; money will be lost; recovery and return to normal operations will not be easy.
- −3 equals significant risk; remedial actions will have to be taken to restore situation; progress toward business goals will be disrupted; effects will be remembered and will have consequences for the future.
- −2 equals some risk, though it will seem to be controllable; within many organizations, would come under "freedom to fail" latitude;

only those most directly affected will understand the setback, which will be considered minor.

- −1 equals very little risk; at worst, will be considered a minor snag; virtually no one will be aware of negative consequences; relatively little work will be necessary to restore status quo.
- 0 equals status quo; performance, morale, image, return, flexibility, and relationships will remain precisely as they are today.
- +1 equals slight improvement; most people will not even realize what has taken place; those directly affected will see a minor advantage; results will be short-lived, and quickly forgotten.
- +2 equals clear improvement; those most closely involved will be highly appreciative of the result; improvement will be tangible and repeatable.
- +3 equals significant improvement; entire organization will either use or be quickly aware of it; effects will be somewhat longer lived; considered a clear advance, which the organization will attempt to exploit.
- +4 equals dramatic improvement; clear competitive inroads and/or operating efficiencies will result; long-term benefits; significant event for organization; will gain outside notice as well.
- +5 equals landmark improvement; turning point and/or watershed event in the development of the company and/or individual; improvement will launch company or department to leadership position; profound change on culture and/or operations of the organization.

Recommendations

Here are some recommendations for determining how to treat opportunities as evaluated on the risk/reward analysis scale and whether or not to carry them forward to the pursuit step. Naturally, these are ultimately your decisions, but these guidelines should be helpful in determining where your opportunities fall.

1. Both best-case and worst-case scenarios rate a positive and are at least +1 or above on the scale: These opportunities should be taken to the pursuit step, without exception.

2. Best- and worst-case scenarios range anywhere from −1 to + 1: These opportunities are seldom worth pursuing because they have minimal impact while requiring attention and focus to see them through correctly. (If they are not followed up on correctly, the negative impact could be much worse than projected.) Sometimes such opportunities can be delegated to others, to test the ability of subordinates to implement. Even in this case, however, you should be

acutely aware of the time and energy investment as compared to the relatively small impact that will result.

3. Best- and worst-case scenarios range from 0 to −5: Obviously, these are not opportunities that should be pursued.

4. In general, when best case is +1 or less and worst case is 0 or worse, do not pursue the opportunity.

5. The best opportunities to pursue are generally those whose positive number exceeds the negative number: for example, a +2 over a −1, a +3 over a −2, a +4 over a −3, a +5 over a −4, or, of course, a better relationship than any of these. A +5 over a −2, for instance, is highly attractive; moving the negative number only two points to the left—creating a +5 over a −4—creates, on the other hand, an opportunity that carries the potential for severe risk. As we've stated earlier, the critical factors come into prominent play when you're dealing with opportunities that bridge the status quo.

6. If you seek to be very safe and conservative in seeking opportunity, never pursue one that rates less than a +2 or less than a −2. That is, the *best* case has to be at least +2, and the *worst* case can be no worse than −2.

7. We recommend that you never pursue an opportunity that rates a −4 or −5 worst case—irrespective of the best case projection—unless you are clear on and confident of the critical factors that need to be addressed to mitigate the worst-case scenario.

Once an opportunity successfully negotiates the development step to your satisfaction, it's time to put the opportunity into practice. We established at the outset of this book that innovation can only be successful when it is implemented. Ideas that aren't implemented aren't innovations. Consequently, we will now turn to the implementation of opportunities, which we call the pursuit step.

Chapter 6

OPPORTUNITY PURSUIT: THE INNOVATOR AS IMPLEMENTOR

Opportunity pursuit is the process of formulating an implementation plan and then beginning the actual implementation of opportunities. It relies heavily on the critical factors that were raised in the development step. It is a step that is specifically designed to focus on individual opportunities and to bridge the gap between the conceptualization and the actualization of these opportunities.

Opportunity pursuit makes innovation happen by allowing you to analyze those factors that will determine success or failure for your opportunity and by assigning *specific* actions that will help to enhance success and avoid failure. Pursuit is begun "on paper"—before energy, resources, and reputations have been committed. Once the plan takes coherent shape, implementation begins. While opportunity pursuit is not an absolute guarantee of success, it does function as an insurance policy, one designed to provide the greatest protection for you and your plan.

There are at least four possible applications of the opportunity pursuit step. First is the pursuit plan becoming the actual implementation plan for a specific opportunity. This usually occurs when you are dealing with an opportunity that is completely within your ability to authorize and implement. In such a case, the pursuit plan is quite detailed and becomes the working document that you and your colleagues use to implement the opportunity.

When an opportunity must fit into a larger planning process (for example, increasing the number of leads generated per salesperson, an opportunity that must fit into a larger marketing plan), the second application of the pursuit plan is to raise the key issues and actions that will become components of that larger plan. Those responsible for pursuing the opportunity are provided with a methodology to ensure that their interests are protected as the larger planning process evolves.

The pursuit plan can also serve as a tactical management tool that enables you to intelligently manage your resources and people. Specific attention is paid to the responsibilities and dates set in the opportunity pursuit step. This tactical plan is reviewed frequently by all concerned in the project to monitor progress and identify emerging trouble spots that were not considered earlier; it also serves as a fail-safe mechanism to safeguard the implementation as much as possible.

The fourth way to use the opportunity pursuit process is in highlighting the fact that key information is missing, or that actions deemed possible in the assessment and development steps are, in fact, not feasible. Although it is not the primary purpose of the pursuit step to serve as yet another filter in the innovation process, we have found that by its very nature of forcing people to identify specific actions and consider specific resources, it can cause last-minute reconsideration of opportunities that are going from "paper" to "action." Perhaps most powerfully, the disciplined, ordered, and systematic process that the opportunity pursuit step provides serves as a very powerful recommending tool because it places the implementation in a visible, objective, and documented format. As an example of the need for such a format, we've observed many fine ideas that were not accepted because:

- superiors couldn't effectively visualize the ideas being presented;
- hackneyed objections were raised that weren't valid for the particular opportunity but couldn't be easily rebutted;
- interests emerged to block the opportunity and carried the day through forcefulness, not through logical argument;
- in the "resale" process, as the opportunity went up the management ranks, succeeding managers were not as effective in selling the idea as were the originators;
- due to financial constraints, new ideas were automatically believed to cost more money, ignoring the key entrepreneurial consideration of *redeployment* of assets and resources to increase yield and productivity;

- reviewers "considered the source" and didn't take the recommendation seriously because they felt those making it simply didn't have sufficient perspective, background, and/or information;
- the recommenders themselves were unorganized, in disagreement about various aspects of the opportunity, and in general, could not present a united, purposeful front.

PREVENTING AND PROMOTING ACTIONS

For an organization, it is as debilitating to ignore good opportunities as it is to implement bad ideas. Consequently, we have found that the pursuit step is especially important in providing a systematic approach to recommendations as well as to implementation. This step is probably most important, however, for encouraging *risk management* and *risk containment.* We've established that innovators take prudent risk, not excessive risk. Opportunity pursuit enables you to objectively determine how much risk is present and what you can realistically do to minimize or eliminate it. Appropriate actions then become a part of the implementation.

There are two basic parts to the opportunity pursuit step. First is the opportunity pursuit worksheet, which is the direct bridge from the development step. The second part is the pursuit plan itself.

For any specific opportunity, review the critical factors that were listed and considered in the development step and list them on the opportunity pursuit worksheet. First examine those that you considered positive, that is, those that will tend to support the best-case outcome you projected for the opportunity. Now begin to generate one or more promoting actions that will help to ensure that each positive critical factor actually occurs in the manner that you desire. For example, if one of the positive critical factors you identified is "the support of the sales director," then a promoting action—an action that will tend to cause this critical factor to occur—might be a private demonstration for the sales director and his or her staff.

In Figure 6–1, you'll see "support of the sales director" listed under Column A, "Positive Factors." You'll see the promoting action, "private demonstration for the sales director and staff," listed below it, in Column C, "Promoting Actions."

Don't feel limited by the numbers of positive factors or by the size of the worksheet. For example, another promoting action for this same critical factor might be to volunteer to help the sales director on her territory management project. You might develop five, six, or even twelve promoting actions for any one critical factor. Obviously,

Figure 6–1. Opportunity Pursuit Worksheet.

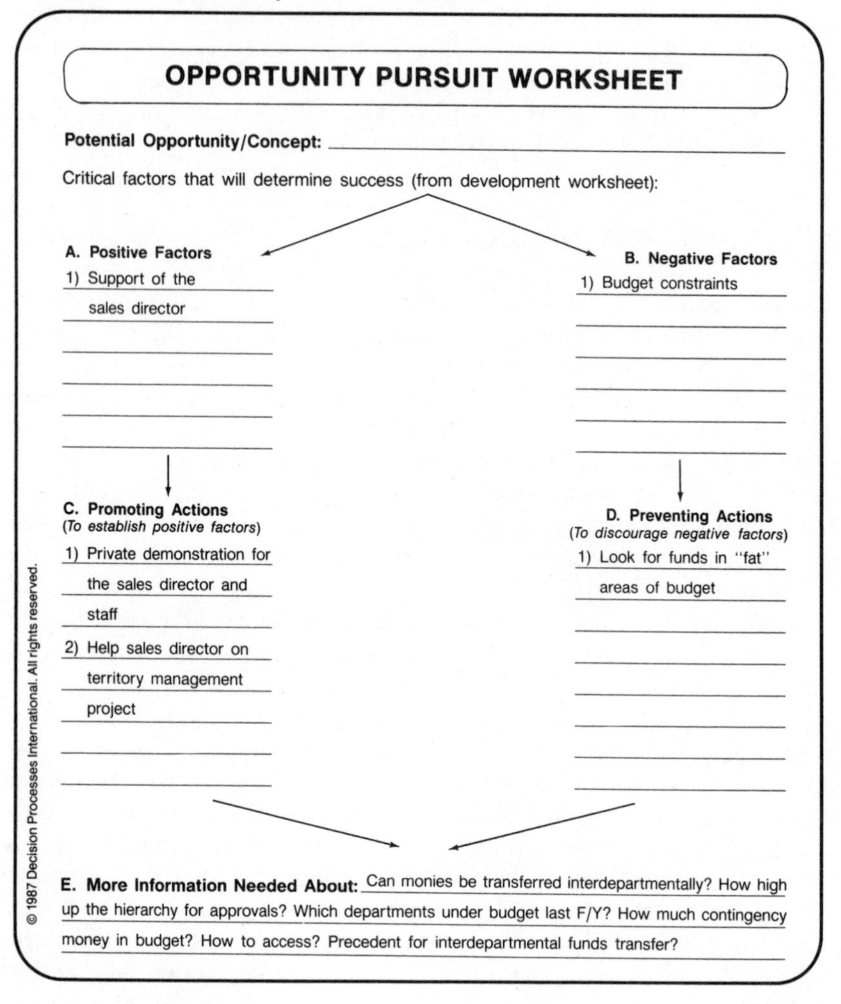

OPPORTUNITY PURSUIT WORKSHEET

Potential Opportunity/Concept: ______

Critical factors that will determine success (from development worksheet):

A. Positive Factors
1) Support of the sales director

B. Negative Factors
1) Budget constraints

C. Promoting Actions
(*To establish positive factors*)
1) Private demonstration for the sales director and staff
2) Help sales director on territory management project

D. Preventing Actions
(*To discourage negative factors*)
1) Look for funds in "fat" areas of budget

E. More Information Needed About: Can monies be transferred interdepartmentally? How high up the hierarchy for approvals? Which departments under budget last F/Y? How much contingency money in budget? How to access? Precedent for interdepartmental funds transfer?

© 1987 Decision Processes International. All rights reserved.

the more promoting actions you develop, the higher the chances that the positive critical factor you want to occur will indeed occur. But when do you have enough? How much time is it worth to spend on developing these? Here are two quick guidelines:

1. The broader the range on the risk/reward analysis from the development step—the more the best case and worst case for the opportunity spans the status quo—the more time you should invest in being very careful in preventing the negative critical

factors. The reason for this is that there is apparently a large downside risk.

2. You know from your intuition, experience, research, and/or perspective that a particular positive critical factor will mean the difference between success and failure. This is often a critical factor that has to do with organizational culture, politics, timing, or key positions in the hierarchy. You will want to surround that critical factor with promoting actions.

You may believe that some positive critical factors will happen with little or no help from you. For those factors, you might choose few or even no promoting actions, depending upon your confidence level. There is no one-to-one relationship between factors and actions. Some critical factors might receive one promoting action, others five or six. It all depends upon your judgment, your perspective, and your knowledge of what has to be done. You should be able to see at this point, however, the importance of raising the critical issues during the best- and worst-case analysis that was performed in the development step. We think that this is a unique step in the innovation process, one that tends to guarantee practical and realistic application of new ideas.

In a similar manner, you should turn to the negative critical factors, those that will tend to produce the worst-case scenario according to your work in the development step. For example, if a negative critical factor is "budget constraint," then a *preventing action* might be an investigation of where money can be found in a "fat" part of the budget, that is, an area that won't produce the kind of productivity and yield gains that your opportunity will. On the worksheet, you'll see "budget constraint" listed under Column B, "Negative Factors," and "look for funds in 'fat' areas of budget" listed beneath that under Column D, "Preventing Actions."

It was only about fifteen years ago that men's hair spray was first introduced into the marketplace. It became very apparent to the manufacturers that the perception of hair spray as a female product was the key critical factor in a worst-case scenario: refusal of the male target market to accept the idea of hair spray as a masculine grooming aid. The preventive action chosen was to use stereotypical "he-men" in massive advertising campaigns to dispel the perception. Long after we've forgotten which product he was pushing, we can all fondly remember Yogi Berra gruffly proclaiming on television that he used hair spray. The success of identifying this critical area and providing effective preventive action can be seen in the fact that men's hair spray has become an integral part of the male toiletries market.

As a result of this first part of the opportunity pursuit step, both positive and negative critical factors from the development step should have been listed (these should be *identical* to the factors raised in the development step), and appropriate promoting and preventing actions should have been assigned to these factors. It's important to do this prior to the actual implementation. Effective actions should be determined prior to considering who does them, when they are done, and how much they will cost. All too often, effective actions are not taken because they tend to conflict with other priorities, personalities become an issue, and organizational bureaucracy tends to get in the way. Once the promoting and preventing actions are determined in this step, they are ready to go into the planning process.

You can always add further promoting and preventing actions as you learn more about the issues and think of better ways to handle the critical factors. One should *never,* however, return to remove promoting and preventing actions for the sake of scheduling, timing, or personalities. If such actions are deemed impossible because of budget constraints or lack of approval, then that is cause for serious deliberation, especially since it may mean that the opportunity itself cannot be successfully implemented. These crucial actions should also never be changed because of minor or petty concerns. It can't be overemphasized that these are the actions that determine the viability of the opportunity; they stem from the critical factors underlying the successful implementation of it. And by *critical* factors, we mean just that: those factors that spell the difference between success and failure. So the first part of opportunity pursuit enables you to analyze how well (practically, financially, politically, culturally, rapidly, etc.) you can minimize risks and exploit benefits. You are required to decide whether the time, money, and effort needed to control risk are justified by the advantages. There are no magic, numeric formulas to accomplish this. Such analysis depends upon your management judgment. The process doesn't replace such judgment, but does allow for a more disciplined, objective evaluation of the facts.

There is one more category on the worksheet. This is Column E, "More Information Needed." As we mentioned earlier, it sometimes becomes apparent in this first step of opportunity pursuit that there is insufficient information available to make determinations about effective promoting and preventing actions. When this happens, the process does provide this additional safeguard, forcing you back to find the information required. At this juncture, it is unwise to proceed with gaps in the framework. Consequently, our preventing action that called for an investigation of other parts of the budget where additional money could be found might require us to seek more information about:

- whether monies can be transferred interdepartmentally;
- how high up the hierarchy budget transfers must be approved;
- which departments came in under budget during the last fiscal year;
- how much money has been planned for contingencies, and how that money can be accessed;
- if there is precedent to cite for such monies being made available.

BUILDING THE PURSUIT PLAN

This brings us to the second part of the opportunity pursuit step, which is the pursuit plan itself. We have emphasized that we favor a simple, uncluttered, and orderly approach to this planning. You may wish, however, to incorporate the planning tools that you have found to be most effective or that your organization requires. We are not suggesting that there is anything unique or ideal about our particular planning format. We do feel, however, that it demonstrates the simplicity with which an opportunity can be planned and implemented. As we move from search, through assessment, through development, and finally to pursuit, we feel that the process should not become more and more complex, especially since at this point the opportunity might well involve people and departments who were not involved from the beginning of the innovation process. Therefore, it is even more essential that the work that has taken place up to this point is configured in a way that makes it readily understandable, easily communicated, and quickly summarized.

As you can see in Figure 6–2, the items listed under "Plan Steps" can be as detailed as you wish. The amount of detail will depend on which of the planning purposes you have in mind: a plan for the implementation of a specific opportunity; a plan for implementation that is to be made part of a larger plan; a tactical, management implementation document; or a guide to develop further information. We've seen many instances where plan steps, when circulated to those responsible, form the basis of a new pursuit plan worksheet for *each* of those individuals. For example, if one step in the plan is "notification to all retail dealers of a new discounting policy," the individual responsible for that step might in turn create substeps and target dates within that step:

Figure 6–2. Opportunity Pursuit Plan.

4 PURSUIT

PURSUIT PLAN

OPPORTUNITY/CONCEPT: Launch New Discounting Policy (sample of evolving plan steps)

PLAN STEPS	TARGET DATE	RESPONSIBILITY
Assemble project team	February 1	Me
Pursue other budget possibilities	March 1	Me, with VP Finance
Research interdepartmental funds transfer	March 1	Administrative Assistant
Private demonstration for sales director and staff	April 15	Project Manager
Help sales director on territory management project	April 15–25	Project Manager
		Sales Manager Staff
Arrange for toll-free dealer inquiry number	May 1	Office Manager
Notify retailers of new discounting policy	June 1	VP Sales and Marketing

© 1987 Decision Processes International. All rights reserved.

- Update list of retail dealers
- Draft document for approval on actual discount policy
- Provide toll-free number for questions from dealers
- Run discounting comparison with that of competitors

If that individual manages other people, he or she is now able to assign responsibilities for these substeps. It is in this manner that a general pursuit plan is turned into individual elements and actions and, most importantly, is assigned to specific people to make sure that things get done. This is the core of effective implementation. Nothing can "fall through the slats" if specific responsibilities are assigned and clear completion dates are agreed on. As progress meetings take place, they should revolve around these pursuit planning worksheets so that attention remains focused on what was accomplished, by whom, on what dates, and what needs to be done in a similar fashion in the future. This allows for adjustments and contingencies to be included in the plan as it evolves. Opportunity pursuit should be an organic and dynamic process, changing as implementation moves forward and adapting to new conditions as necessary.

The two elements that *must* be a part of the plan steps, however, are the promoting and preventing actions from the previous worksheet. This is the finalization of the bridge from the development step to implementation in that critical factors are being addressed through the inclusion of their appropriate promoting and preventing actions as a part of the planning process. From our earlier example, a private demonstration for the sales director and staff must be included as one of the plan steps, with a target date and responsibility assigned. Similarly, the investigation of fat parts of the budget has to be included as a plan step, and also assigned. These actions may be the heart of the implementation process, because they are the key to ensuring the critical factors that will tend to promote success and to mitigating or eliminating the critical factors that will tend to hinder success. Consequently, if you see target dates missed, responsibilities shirked, and progress generally missing at key checkpoints, you know immediately that you must regroup—implementation is going awry. Without these important actions being carefully considered and included in the plan, the best innovations—despite their relative benefit and appeal—can become nothing more than the roll of the dice that we tried to avoid from the outset of the innovation process.

Chapter 7
SYSTEMATIC INNOVATION IN OPERATION

You've come to the final chapter in another book. You're in a position that you've probably been in many times before. You've read some good ideas, considered some practical examples. It's hard to see how you could fail to use in the near future what you've learned. In fact, as soon as that right situation presents itself, you'll be sure to apply what you've learned in this book.

Perhaps that's exactly what will happen, but the likelihood is that unless you take some very specific steps immediately, you'll use precious little of what you've learned. Unhappily, our experience is that most people tend to be excellent *acquirers* of information, but very poor *appliers* of information to achieve practical results.

In the long run, we feel that it all boils down to a question of hard, disciplined mental work. Those who are willing to consciously try to use the process—and learn from their failures and gain from their successes in so doing—are those who will master it and use it most effectively in the future. Just as innovation isn't complete until opportunities are *implemented,* your gain from this book won't be complete until the techniques and processes you've learned have been *applied.*

There are two aspects to successful application. One is the effective use of the innovation process by the individual; the second is an organizational application that allows such use to achieve its maximum potential. We will examine each of these basic components in turn.

INDIVIDUAL SUCCESS

Here are the keys we've discovered to effective individual use of the processes and approaches described in this book. The more you're able to employ these tactics in your day-to-day routines, the more effective an innovator you'll be.

Focus on improvement, not just on fixes. Whenever you find yourself called upon to solve a problem, spend some time investigating whether the situation can be improved and not just fixed. Whenever the mental set is, "We've got a problem that we have to fix," the best you can hope for is a return to the previous performance level. But when the attitude is, "Let's fix this problem *and* decide how to improve the situation," you have the proper momentum established for effective innovation.

Remove the phrase "Yes, but . . ." *from your vocabulary, and don't let others use it in your presence.* If you're ever curious about just how much of a dampening effect this attitude has on innovation, count the number of times you hear the phrase used in a meeting, or during any workday. You'll be shocked at the number of times it's used and by the number of people using it. Instead, allow ideas to breathe and to grow. There will be plenty of time later, in the assessment and development steps of the process, to critically analyze ideas and discard those that are impractical. But at the outset, make sure your new ideas get on the table, and allow others to establish theirs as well.

Keep a list of the ten search areas (and any other areas you might have added to them) *close by and visible where you perform most of your work.* We have found that by simply keeping this list nearby as a memory-jogger, people are more apt to view the environment in terms of opportunity. Thinking about the search areas will eventually become second nature, but as we stated earlier, it's the initial discipline in using them that's key to successful application.

Use the impact scale and risk/reward analysis from the development step to calculate prudent risk. Whether you use our calibrations or some others of your own, utilize some objective scale to determine the amount of risk present in any given opportunity. If best- and worst-case projections are both superior to the status quo, you know you probably have a winner. If both are inferior to the status quo, you know you probably have a loser. But more importantly, the degree to which best- and worst-case projections may *span* the status quo will tell you the degree of risk that you are exposing yourself to and will highlight the need to establish effective actions to mitigate that risk. Risk should not be viewed as an excuse to do nothing. Rather, it should be viewed as a management challenge to *do* something. When risk is excessive, you have good reason not to act. But you must be able to demonstrate why the risk can't be managed. The reason that people choose diversified stock portfolios or mutual funds, or choose to purchase insurance that they hope they'll never have to use, is in order to better manage risk. If the premium is greater than the amount at risk—greater than the value of your investment—you obviously won't take out the insurance.

Before moving ahead with implementation of an opportunity, ascertain the critical factors that will determine success or failure. Most people overlook this step. There are almost always a few critical factors that must be managed if an opportunity—or for that matter, any decision—is to be successfully implemented. Only by identifying those critical factors will you be in a position to take effective promoting and preventive actions. Remember, effective implementation is the linchpin of innovation. Identification of the critical factors and determining the appropriate actions to manage them is the crux of implementation.

Never assume that an opportunity is going to implement itself. Whether large or small, an opportunity must be guided into implementation by people who have specific responsibility for so doing.

ORGANIZATIONAL SUCCESS

Of course, the organization can and should contribute to individual efforts by stimulating, encouraging, and sustaining innovation in its midst.

Peter Drucker observed in his book, *The Effective Executive,* "An organization is not, like an animal, an end in itself, and successful by the mere act of perpetuating the species. An organization is an organ of society and fulfills itself by the contribution it makes to the outside environment."[1] All over the world, those organizations that have tended to make the biggest contributions to the outside environment, and to society in general, have been those that have been most innovative. Innovation is not culturally bound. It takes place in a variety of societies, countries, and conditions. Nor is innovation a function of a particular pursuit, industry, product, or service. We have found innovation in industries as "old-line" as the steel industry (specialty steel makers using high-technology operations), as traditional as the fur industry (furriers providing low-cost "bridge" furs purchased by young working women for their own use), and as common as auto care (companies that will pick up your car at your door, service and clean it, and return it to you that evening).

Nor is innovation determined by organization culture. While the popular business press often takes the position that organizations must radically change their culture before they can foster innovation—and that such efforts take several years' worth of preparation until they yield results—we have found that making the culture conducive to innovation is a straightforward task that can be effective almost immediately. The most direct route to creating a culture that's amenable to innovation is simply to stop saying "NO!" to enterprising employees who present new ideas.

Why do managers, even those committed to promoting innovation, regularly reject new ideas? The answer is not that the culture is "wrong." Rather, it is that the individuals presenting the ideas often cannot explain the rationale behind their recommendation or they have difficulty reconstructing the thinking process they used. Or, sometimes, the thinking process has been short-circuited, resulting in weak or incomplete ideas.

To promote innovation within organizations, don't get derailed by elaborate culture-change programs; instead, invest your time and efforts far more wisely by providing your people with a systematic and visible process of innovation that enables them to identify innovative business opportunities, assess them against other options, explore their potential rewards and risks, and then develop plans for pursuing the best ones.

Finally, even in those organizations that seem to be driven by the force of a charismatic innovator at the top—Victor Kiam at Remington, former Mayor Schaeffer of the city of Baltimore, Akio Morita at Sony, Bill Gates at Microsoft—innovation is perpetuated by an attitude that encourages people *throughout* the organization to emulate the innovative directions of the top person. The larger the organization, the more innovation has to be a ground swell movement. An individual can be the sole innovator in running a three-person print shop or a ten-person travel agency, but not in running a $50-million mail-order business, or Motorola.

Here's a quick test you can use to assess the entrepreneurial spirit and innovative leanings of your organization. After each question, decide whether your organization rates a "one," "two," or "three." A "one" represents "infrequently or never," a "two" represents "sometimes," and a "three" represents "usually or always." Use your best judgment in answering the questions. If you're uncertain about something, it's probably safe to assign it a "two."

1. Our organization is seen by its customers as being on the leading edge of its field.

2. Our organization is usually the one that the competition is trying to catch up with.
3. We allow people the "freedom to fail" and give careful consideration to new ideas, no matter what their origins.
4. Innovative people in our organization are held up as examples and are clearly recognized by senior management for their contributions.
5. Formal time is provided in meetings for examining opportunities, time that is quite distinct and apart from time spent solving problems and looking at existing operations.
6. We tend to hire people for their talent, welcoming diversity, and don't attempt to hire people all cut from the same mold.
7. We look at seemingly unrelated events in the environment to determine how they might benefit us, our products and services, and/or our customers.
8. Our company culture tends to look at change as containing opportunity, not threat.
9. We are methodical about innovation, particularly in utilizing processes to assess the relative value of new ideas that come before us.
10. We have used our mistakes in the areas of new products and/or improvements to the organization to enhance similar ideas that have arisen later.
11. Our organization, both line and staff, tends to get excited about new developments, new ideas, and new client approaches.
12. We have utilized formal training sessions to help people free up their creative energies and become more consciously innovative.
13. Our company's stated strategy includes the need to be innovative and to explore new approaches in the future.
14. My immediate colleagues present a good sounding board for new ideas and are not hesitant about generating new approaches and new ways of doing things.
15. Even smooth-running aspects of the operation are evaluated to see if the return can be improved or if the investment would be better made elsewhere.
16. The organization discourages the attitude, "You won't get rich, but you won't get fired," and there is little deadwood to be found within key management positions.
17. Rules and standard operating procedures are sometimes broken when there seems to be the opportunity to achieve a breakthrough or a new level of performance.

18. In their oral and written messages to me and my colleagues, our superiors cite the need to be innovative, entrepreneurial, and creative.
19. Our organization has the history and/or capability of achieving dramatic breakthroughs in growth, market penetration, service quality, and/or product development.
20. Articles, war stories, and examples of innovation in other firms and other industries are the topic of conversation in our organization, both formally and informally.

Add up your ratings, and score as follows:

20–29: Your organization is decidedly uninnovative, and is probably geared toward frustrating such efforts.

30–39: Your organization tends to be slothful about innovation and is able to achieve it only through the efforts of forceful personalities or through absolute market demand.

40–49: Your organization is situationally innovative. This means that there are repositories throughout the organization of innovative thinking and action. It's more a matter of luck, however, than of design.

50–60: Yours is a highly innovative organization, with procedures and techniques in place to foster, stimulate, and reward creativity. Although the personality of the CEO and other top officers may be largely responsible, the probability is that such innovative actions are institutionalized and perpetuated by the culture of the organization itself.

What do organizations do to encourage true innovation from within their ranks? We have identified the following characteristics of innovative organizations. The list isn't meant to be exhaustive, nor do all traits apply to all innovative companies. These characteristics, however, do tend to be generic and are usually to be found within any successfully innovative organization.

Traits of Innovative Organizations

Innovative organizations are close to their customers. Innovative companies listen to their customers and clients. This doesn't mean they simply open mail and listen to complaints. They proactively solicit customer opinions and seek customer input on marketing strategies, product design, and service levels so as to continually *improve* products and services.

Innovative organizations are value-driven; innovation is one of the essential values. These are organizations that have embraced innovation beyond the bounds of lip service and superficial mentions in elaborate strategy statements. Innovation is exemplified by top management, and people are encouraged to experiment, create, and innovate. The "freedom to fail" is taken for granted, and prudent risk-taking is encouraged. By definition, company values should be taken into account in every decision and plan that the company considers. Consequently, innovation becomes a fundamental aspect of the company's decision making and planning.

Innovative organizations provide entrepreneurs working for them with a measure of autonomy. They are flexible enough to realize that rules and procedures need to be bent, or even broken, for successful innovation to take place. Intrinsic to this approach is the flexibility necessitated by rapidly changing business conditions and environments. When IBM decided to enter the personal computer marketplace, it was smart enough to realize that the new product could never be developed within its traditional framework. Consequently, an autonomous group was created to oversee the creation of the IBM-PC. The top managers in this autonomous group had a clear and key charter: protect the group from the traditional systems, procedures, and regimentation of the IBM system. They were able to effectively accomplish this, but only because the organization recognized at the outset the need for such safeguards.

Innovative organizations make planning an exercise in creativity rather than confine it to simple forecasting. All too often planning is nothing more than a straight-line extrapolation of where you are today projected into the future. This automatically limits what can be accomplished because it makes assumptions based upon today's realities and current growth. Strategic planning needs to be separated from operational planning. By this we mean that strategic planning should be painting a picture—a vision—of what the organization should look like in the future. Therefore, top management should be forced to be innovative and creative in seeking opportunities to achieve that future vision. In so doing, the planning process should become totally involved with rapid and early recognition of changes that can be turned into competitive advantage. The planning process should never be defensively oriented—it should be a potent offensive weapon for the organization.

Competitive advantage through innovation is sought throughout an innovative organization, not simply in "major areas." That is, innovation to achieve competitive advantage is more often a function of many "small" things spread throughout the organization than it is of one large breakthrough in sales and marketing. For example, encouraging and rewarding employees who develop better ways to deliver products on time, handle customer inquiries, provide instruction material, and provide warranty service are, cumulatively, probably more effective actions in gaining competitive advantage than any other single action, no matter how large, elsewhere. Organizations do not have to radically reshape themselves to be innovative. They can readily become more innovative simply by using the human resources already in place. The use of these resources should include facilitators, throughout the organization, to serve as the driving forces for the innovation process.

Consistent and incremental improvement is among the most effective kinds of innovation for an organization. The unique breakthrough that dramatically advances technology doesn't occur very often, sometimes not even within one's career in a particular industry. Innovative companies generally try to stay ahead of the competition by constantly seeking small improvements in every aspect of their business. And we mean literally that—*every* aspect. While areas like sales or product design might be the sexiest, it is areas like purchasing, finance, legal, and client service that can provide tremendous return on the innovative investment. Often, a key aspect of innovation is "getting there firstest with the mostest." Even a slight edge in timing can mean a tremendous amount of difference in the marketplace. American Airlines' Advantage Program is the most heavily subscribed-to frequent-flyer program today, not because it offers the best benefits—its benefits are almost identical to those of its competitors'—but because it was the *first* to offer such a program in the marketplace. Arriving first creates a temporary monopoly and a unique market position that may never have to be surrendered. And getting there first is usually a function of innovating in a wide variety of areas that provide the company with the flexibility and agility to innovate slightly faster than the competition.

Here's a quick and simple diagnostic test. Look at the budget for your department, your division, or your position. Are there monies allocated specifically for innovation? That's right, is there any budget provision that allows for the experimentation and investigation necessary for successful innovation? If not, the organization probably isn't doing everything it can to foster such across-the-board innovation. Incremental improvement often goes unnoticed, but it won't occur by itself. It's the result of deliberate stimulation by the organization, including provision for financial resources and the freedom necessary to constantly examine all aspects of the operation and determine how they can be improved. Improvements in *reaction* to customer complaints or competitive inroads are not innovative improvements. And they will seldom create a unique market advantage for the organization.

Organization teams that hold "seek and search" meetings for opportunity and that ask, "What can we improve?" are probably *more* important than problem-solving efforts.

It's been said that there's nothing new under the sun, but we should all know better. One of the greatest sources of "newness" and opportunities is the recombination of existing or "old" ideas. The whole is often greater than the sum of its parts, and *innovative organizations achieve tremendous value by recombining existing approaches, ideas, and methodologies.* Merrill Lynch achieved a unique market position with its cash management account because it was the first of its kind in the market. Yet, what Merrill Lynch did was to combine existing ideas: traditional banking and brokerage services combined with checking, savings, and brokerage accounts under one comprehensive relationship. This provided both the reality and the perception of tremendously increased value for the customer, "merely" through a recombination of existing services.

Recombination and repackaging of products and services should be pursuits that every organization looks at with diligence—these are key sources of innovation. This seems to be a uniquely organizational responsibility; most individuals are not in a position to effect such combinations personally. It takes an organizational response and organizational resources to investigate and execute such recombinations.

Of course, seeing such opportunities in recombination entails the dismantling of traditional taboos. Separate departments, product managers, geographic markets, and traditional "fiefdoms" must cooperate. Among other things, this requires a "win/win" attitude—one that does not thrust internal departments into competition with each other but rather, unites them in a common competition against rival firms. Organizations can encourage such cooperation by scheduling meetings and task forces that are interdivisional in their nature and in their goals. Cross-operational "opportunity sessions" should be held just as often as their cousins, the problem-solving sessions. Rotating executives and key managers among even disparate operations helps to break down the walls still further.

Successful organizations create and maintain realistic expectations. The pursuit of every opportunity can be no more automatically successful than the pursuit of every new client or product. There will be failures. This should not dissuade the organization from innovating. On the contrary, such setbacks should serve to further educate the organization about how best to innovate within its industry and marketplace. Not all of Merck's drugs have turned out to be commercially viable. A relatively small portion of the research at Bell Labs has resulted in scientific breakthrough. Zap Mail was an idea whose time had not come for Federal Express. Beta-VCR technology is rapidly entering the "twilight zone" for Sony. Mercedes-Benz has had to endure an auto recall or two. Nevertheless, these have been truly innovative organizations, ones that stand head and shoulders above their competitors.

Perhaps the true mark of an innovator is in suffering such setbacks. If no such setbacks are ever endured, one could make the argument that the organization has simply eschewed involvement in true innovation. Innovation does involve risk, albeit prudent risk. Risk can never be completely controlled, so some failure is inevitable. Yet, success is never final and failure is seldom fatal—it's courage that counts. We mean the courage of an organization to stick by its belief in the value of innovation to better serve its customers and to better enhance the value of business. Such a belief—or value—is not easily shaken by temporary setback or isolated failure.

Successful organizations measure innovation's progress against specific targets, such as 3M's mandate for generating "25 percent of sales from new products developed over the prior five years." Unless you establish realistic and clear goals, as you would for any business endeavor, how can you measure your progress?

Innovative organizations charge specific people with the responsibility for anticipating change. These are the lookouts, high above the pitching deck, with a unique view of the horizon. Only by recognizing change at this distance can the course of the organization be effectively adapted and the change exploited.

We've talked about the role of charismatic leaders as the visionaries most involved with change. Yet, this needs to be a role played by a great diversity of people within the organization. We've discussed the role of the strategic planners and the importance of their accounting for change in the planning process, as well as the need for them to be offensively minded about change, not defensively minded. Yet, we believe that an organization can empower *everyone* to be sensitive to anticipating, recognizing, and exploiting change.

In fact, we believe that the greatest responsibility lies with the line manager. This is the person most in contact with the customer, the company's products and services, and the support mechanisms of the organization. The line manager is actually the "point person" in the search for change. The more an organization encourages line management to seek out change, explore innovative opportunities, and take prudent risks, the more that organization is likely to set an effective course through the tides of change.

REPEATABLE BUSINESS PRACTICE

Our observations have demonstrated to us that both individuals' initiatives and a receptive organization are important to successful innovation. No matter how receptive the organization, however, successful innovation will not take place unless individuals are equipped with the skills, talents, and processes that allow innovation to occur. Innovation will not spring up spontaneously—it takes individuals who are willing and able to start the process.

Conversely, no matter how hostile the environment—no matter how unfriendly the organization may appear—individuals with the skills and process and determination to innovate will still be able to do so. So while the optimal combination is individual skills coupled with a receptive organization, *it is the individual skills that are the most important.* That's why we found individuals able to implement in all types of organizations, at any level of the hierarchy, and in any number of circumstances. Whether or not the organization gives them any help, they are able to help themselves.

A fundamental reason why these individual skills are so important is that they are *transferrable.* They remain with the individual through transfers, promotions, and changes in environment. And they can be communicated to others in a rational and methodical manner. While we have provided the model for the entire innovation process in these pages, we hope that it's clear at this point that any bits and pieces of the model can be applied situationally. If you wait for just the "right moment" to take out your charts and use the system from A to Z, the chances are that you will achieve very few practical applications. On the other hand, if you're willing to use the process situationally and flexibly, realizing that 80 percent of its use is going to be mentally performed and quickly done, you'll soon find yourself innovating on a daily basis. It's important to understand the whole before you can use the parts, but once you do, the parts can be used as necessary.

So don't wait for that perfect opportunity to innovate. That's what everyone else is waiting for. You should be taking *every* opportunity, no matter how imperfect, to innovate. That's what few people realize, and that's what will put you ahead of the pack.

We've attempted to stress throughout this book the importance of being innovative on both an individual and an organizational level. We've provided some basic, sequential, and disciplined steps that constitute a *process* that can be used as a repeatable business practice to achieve innovation.

Innovation is the result of focused, disciplined, and rigorous mental work. There is no shortcut around such work. But the process we've provided is a structured way in which to accomplish your entrepreneurial goals. Like any mental process, the more one uses this process, the better one gets at it; and the better one gets at it, the more results are achieved. The more results are achieved, the less "work" seems to be involved in the process.

How will you know when you're innovating well? Apart from the results that you generate, you'll know you're innovating well when you no longer think about "applying the process," or about "using the search step." You're an effective innovator when innovation is the result of doing what comes naturally—because the process has been assimilated into your work habits and, one would hope, inculcated into the organizational norms.

No one personality type is the best and most effective at innovating. It is only a question of a willingness to look at things differently, to view change as opportunity, and to be constantly seeking *improvement.* Perhaps there *is* nothing new under the sun. Innovative techniques have been in use since time immemorial, manifested from the pyramids of the Egyptians to the gunpowder of the Chinese. Individuals who have best achieved their personal goals and contributed the most to society and their organizations are those who have been consistently innovative. We hope you'll begin to use these techniques as soon as you turn this page, to begin to better achieve your own and your organization's goals immediately.

Do not seek to follow in the footsteps of the men of old; seek what they sought.

—Matsuo Basho (1644–1694)

NOTE

1. Peter Drucker, *The Effective Executive* (New York: Harper & Row, 1966).

Index

About the Authors

Michel Robert is the founder and president of Decision Processes International, Inc., a firm established in 1980 that specializes in the processes that lead to superior management performance and employee productivity. DPI's clients include 3M, Fiat, Caterpillar Tractor, Noranda Group, Digital Equipment, Philips, and Northern Telecom. Mr. Robert has worked with clients all over the world, publishes frequently, and is the author of the acclaimed book, *The Strategist CEO: How Visionary Executives Build Organizations.* He is personally in demand to work with top executives of major organizations in establishing the nature and direction of their businesses. He has served in senior positions with corporations that include Johnson & Johnson, Nabisco Brands and Smith, Kline and Beckman. Mr. Robert currently resides in Weston, Connecticut with his wife Ellie.

Alan Weiss is the founder and president of the Summit Consulting Group, Inc., a firm established in 1983 that focuses on critical areas of organization development. The firm's clients include Merck & Company (recently voted "America's most admired company" for the second consecutive year), IBM, GTE, Marine Midland Bank, Pillsbury, City University of New York, and the U.S. Department of Justice. Mr. Weiss has published over 200 articles on organization and management development, has contributed to two books, and appears as a featured speaker all over the world. He is former president of Kepner-Tregoe Continuing Education of Princeton, New Jersey and of Walter V. Clarke Associates of Providence, Rhode Island and has managed profit centers in North America, South America, and Asia. *Success* magazine has described him as "a worldwide expert in trends in executive education." Since 1986 he has also served as a partner in Decision Processes International, responsible for program and process research and development. He resides with his wife and children in East Greenwich, Rhode Island.

can "fall through the slats" if specific responsibilities are assigned and clear completion dates are agreed on. As progress meetings take place, they should revolve around these pursuit planning worksheets so that attention remains focused on what was accomplished, by whom, on what dates, and what needs to be done in a similar fashion in the future. This allows for adjustments and contingencies to be included in the plan as it evolves. Opportunity pursuit should be an organic and dynamic process, changing as implementation moves forward and adapting to new conditions as necessary.

The two elements that *must* be a part of the plan steps, however, are the promoting and preventing actions from the previous worksheet. This is the finalization of the bridge from the development step to implementation in that critical factors are being addressed through the inclusion of their appropriate promoting and preventing actions as a part of the planning process. From our earlier example, a private demonstration for the sales director and staff must be included as one of the plan steps, with a target date and responsibility assigned. Similarly, the investigation of fat parts of the budget has to be included as a plan step, and also assigned. These actions may be the heart of the implementation process, because they are the key to ensuring the critical factors that will tend to promote success and to mitigating or eliminating the critical factors that will tend to hinder success. Consequently, if you see target dates missed, responsibilities shirked, and progress generally missing at key checkpoints, you know immediately that you must regroup—implementation is going awry. Without these important actions being carefully considered and included in the plan, the best innovations—despite their relative benefit and appeal—can become nothing more than the roll of the dice that we tried to avoid from the outset of the innovation process.